MECHANICAL ENGINEERING

QUICK REFERENCE CARDS

Michael R. Lindeburg, P.E.

PROFESSIONAL PUBLICATIONS, INC.
Belmont, CA 94002

In the ENGINEERING REVIEW MANUAL SERIES

Engineer-In-Training Review Manual
  Engineering Fundamentals Quick Reference Cards
  Mini-Exams for the E-I-T Exam
  1001 Solved Engineering Fundamentals Problems
  E-I-T Review: A Study Guide
Civil Engineering Reference Manual
  Civil Engineering Quick Reference Cards
  Civil Engineering Sample Examination
  Civil Engineering Review Course on Cassettes
  Seismic Design for the Civil P.E. Exam
  Timber Design for the Civil P.E. Exam
Structural Engineering Practice Problem Manual
Mechanical Engineering Reference Manual
  Mechanical Engineering Quick Reference Cards
  Mechanical Engineering Sample Examination
  101 Solved Mechanical Engineering Problems
  Mechanical Engineering Review Course on Cassettes
  Consolidated Gas Dynamics Tables
Electrical Engineering Reference Manual
  Electrical Engineering Sample Examination
Chemical Engineering Reference Manual
  Chemical Engineering Practice Exam Set
Land Surveyor Reference Manual
Metallurgical Engineering Practice Problem Manual
Petroleum Engineering Practice Problem Manual
Expanded Interest Tables
Engineering Law, Design Liability, and Professional Ethics
Engineering Unit Conversions

In the ENGINEERING CAREER ADVANCEMENT SERIES

How to Become a Professional Engineer
The Expert Witness Handbook—A Guide for Engineers
Getting Started as a Consulting Engineer
Intellectual Property Protection—A Guide for Engineers
E-I-T/P.E. Course Coordinator's Handbook
Becoming a Professional Engineer

Distributed by: Professional Publications, Inc.
1250 Fifth Avenue
Department 77
Belmont, CA 94002
(415) 593-9119

MECHANICAL ENGINEERING QUICK REFERENCE CARDS

Printed in the United States of America

ISBN: 0-932276-45-8

Professional Publications, Inc.
1250 Fifth Avenue, Belmont, CA 94002

Current printing of this edition (last number): 7 6 5

FOR INSTANT RECALL

CONVERSIONS

multiply	by	to get
atmospheres	29.92	inches of mercury
psi	27.7	inches of water
BTU's	778	ft-lbf
cubic feet	7.48	gallons
gallons	.1337	cubic feet
gpm	.002228	cubic feet/sec
horsepower	33,000	ft-lbf/min
	550	ft-lbf/sec
	746	watts
	2545	BTU/hr
kilowatts	737.6	ft-lbf/sec
	3413	BTU/hr
microns	EE-6	meters
mph	1.467	ft/sec

TEMPERATURE CONVERSIONS

$$°F = 32 + \left(\tfrac{9}{5}\right)°C$$

$$°C = \tfrac{5}{9}(°F - 32)$$

$$°R = °F + 460$$

$$°K = °C + 273$$

$$\Delta°R = \left(\tfrac{9}{5}\right)\Delta°K$$

$$\Delta°K = \left(\tfrac{5}{9}\right)\Delta°R$$

SI CONVERSIONS

multiply	by	to get
kPa	0.145	psi
kg	0.068522	slug
N	0.2248	lbf

DERIVED DATA, PHYSICAL PROPERTIES, AND UNITS

atmospheric pressure: 14.7 psia, 29.92 inches of mercury, 760 torr, 1 bar

density of air at 1 atmosphere and 70°F: 0.075 lbm/ft^3

density of mercury: 0.491 lbm/in^3

density of water: 62.4 lbm/ft^3, 0.0361 lbm/in^3, 1 g/cm^3

standard gravity: 32.2 ft/sec^2, 386 in/sec^2

modulus of elasticity for steel: 3 EE7 psi (hard), 2.9 EE7 psi (soft)

modulus of shear for steel: 1.2 EE7 psi

Poisson's ratio for steel: 0.3

molecular weight of air		29.0
	carbon	12.0
	carbon dioxide	44.0
	helium	4.0
	hydrogen	2.0
	nitrogen	28.0
	oxygen	32.0

ratio of specific heats for air: 1.4

specific gas constant for air: 53.3 ft-lbf/°R-lbm

specific gravity for mercury		13.6
	water	1.0

solar constant: 429 $BTU/hr\text{-}ft^2$

specific heat of ice		0.5 BTU/lbm (approx.)
	water	1.0
	steam	0.5 (approx.)
	air	0.24 (constant pressure)

heats of combustion for hydrogen		60,958 BTU/lbm (high)
	gasoline	20,260
	no. 1 diesel	19,240
	ethyl alcohol	12,800
	carbon	14,093
	sulfur	3,983
	coal	12,000 (approx.)
typical film coefficients for still air		1.65 $BTU/hr\text{-}ft^2\text{-}°F$
	moving air	6.0
	steam (condensing)	2000

FORMULAS

area of a circle πr^2

circumference of a circle $2\pi r$

area of a triangle

$$\tfrac{1}{2}bh$$

volume of a sphere

$$\left(\tfrac{4}{3}\right)\pi r^3$$

area moment of inertia for a rectangle

$$I_{centroid} = \frac{bh^3}{12}$$

$$I_{side} = \frac{bh^3}{3}$$

area moment of inertia for a circle

$$I_{centroid} = \frac{\pi r^4}{4}$$

polar area moment of inertia for a circle

$$J = \tfrac{1}{2}\pi r^4$$

mass moment of inertia of cylinder

$$J = \tfrac{1}{2}mr^2$$

pressure and head

$$p = \rho h$$

rotational speed

$$\omega = 2\pi f = 2\pi\left(\frac{rpm}{60}\right)$$

decibels

$$dB = 10\log_{10}\left(\frac{P1}{P2}\right)$$

COMPUTATIONAL VALUES OF FUNDAMENTAL CONSTANTS

Constant	SI	English
charge on electron	−1.602 EE−19 C	
charge on proton	+1.602 EE−19 C	
atomic mass unit	1.66 EE−27 kg	
electron rest mass	9.11 EE−31 kg	
proton rest mass	1.673 EE−27 kg	
neutron rest mass	1.675 EE−27 kg	
earth weight		1.32 EE25 lb
earth mass	6.00 EE24 kg	4.11 EE23 slug
mean earth radius	6.37 EE3 km	2.09 EE7 ft
mean earth density	5.52 EE3 kg/m^3	3.45 lbm/ft^3
earth escape velocity	1.12 EE4 m/s	3.67 EE4 ft/sec
distance from sun	1.49 EE11 m	4.89 EE11 ft
Boltzmann constant	1.381 EE−23 J/°K	5.65 EE−24 $\frac{ft-lbf}{°R}$
permeability of a vacuum	1.257 EE−6 H/m	
permittivity of a vacuum	8.854 EE−12 F/m	
Planck constant	6.626 EE−34 J·s	
Avogadro's number	6.022 EE23 molecules/gmole	2.73 EE26 $\frac{molecules}{pmole}$
Faraday's constant	9.648 EE4 C/gmole	
Stefan-Boltzmann constant	5.670 EE−8 W/m^2-K^4	1.71 EE−9 $\frac{BTU}{ft^2-hr-°R^4}$
gravitational constant (G)	6.672 EE−11 m^3/s^2-kg	3.44 EE−8 $\frac{ft^4}{lbf-sec^4}$
universal gas constant	8.314 J/°K−gmole	1545 $\frac{ft-lbf}{°R-pmole}$
speed of light	3.00 EE8 m/s	9.84 EE8 ft/sec
speed of sound, air, STP	3.31 EE2 m/s	1.09 EE3 ft/sec
speed of sound, air, 70°F, one atmosphere	3.44 EE2 m/s	1.13 EE3 ft/sec
standard atmosphere	1.013 EE7 N/m^2	14.7 psia
standard temperature	0°C	32°F
molar ideal gas volume (STP)	22.4138 EE−3 m^3/gmole	359 ft^3/pmole
standard water density	1 EE3 kg/m^3	62.4 lbm/ft^3
air density, STP	1.29 kg/m^3	8.05 EE−2 lbm/ft^3
air density, 70°F, 1 atm	1.20 kg/m^3	7.49 EE−2 lbm/ft^3
mercury density	1.360 EE4 kg/m^3	8.49 EE2 lbm/ft^3
gravity on moon	1.67 m/s^2	5.47 ft/sec^2
gravity on earth	9.81 m/s^2	32.17 ft/sec^2

ATOMIC WEIGHTS OF ELEMENTS REFERRED TO CARBON (12)

Element	Symbol	Atomic Weight	Element	Symbol	Atomic Weight
Actinium	Ac	(227)	Mercury	Hg	200.59
Aluminum	Al	26.9815	Molybdenum	Mo	95.94
Americium	Am	(243)	Neodymium	Nd	144.24
Antimony	Sb	121.75	Neon	Ne	20.183
Argon	Ar	39.948	Neptunium	Np	(237)
Arsenic	As	74.9216	Nickel	Ni	58.71
Astatine	At	(210)	Niobium	Nb	92.906
Barium	Ba	137.34	Nitrogen	N	14.0067
Berkelium	Bk	(249)	Osmium	Os	190.2
Beryllium	Be	9.0122	Oxygen	O	15.9994
Bismuth	Bi	208.980	Palladium	Pd	106.4
Boron	B	10.811	Phosphorus	P	30.9738
Bromine	Br	79.909	Platinum	Pt	195.09
Cadmium	Cd	112.40	Plutonium	Pu	(242)
Calcium	Ca	40.08	Polonium	Po	(210)
Californium	Cf	(251)	Potassium	K	39.102
Carbon	C	12.01115	Praseodymium	Pr	140.907
Cerium	Ce	140.12	Promethium	Pm	(145)
Cesium	Cs	132.905	Protactinium	Pa	(231)
Chlorine	Cl	35.453	Radium	Ra	(226)
Chromium	Cr	51.996	Radon	Rn	(222)
Cobalt	Co	58.9332	Rhenium	Re	186.2
Copper	Cu	63.54	Rhodium	Rh	102.905
Curium	Cm	(247)	Rubidium	Rb	85.47
Dysprosium	Dy	162.50	Ruthenium	Ru	101.07
Einsteinium	Es	(254)	Samarium	Sm	150.35
Erbium	Er	167.26	Scandium	Sc	44.956
Europium	Eu	151.96	Selenium	Se	78.96
Fermium	Fm	(253)	Silicon	Si	28.086
Fluorine	F	18.9984	Silver	Ag	107.870
Francium	Fr	(223)	Sodium	Na	22.9898
Gadolinium	Gd	157.25	Strontium	Sr	87.62
Gallium	Ga	69.72	Sulfur	S	32.064
Germanium	Ge	72.59	Tantalum	Ta	180.948
Gold	Au	196.967	Technetium	Tc	(99)
Hafnium	Hf	178.49	Tellurium	Te	127.60
Helium	He	4.0026	Terbium	Tb	158.924
Holmium	Ho	164.930	Thallium	Tl	204.37
Hydrogen	H	1.00797	Thorium	Th	232.038
Indium	In	114.82	Thulium	Tm	168.934
Iodine	I	126.9044	Tin	Sn	118.69
Iridium	Ir	192.2	Titanium	Ti	47.90
Iron	Fe	55.847	Tungsten	W	183.85
Krypton	Kr	83.80	Uranium	U	238.03
Lanthanum	La	138.91	Vanadium	V	50.942
Lead	Pb	207.19	Xenon	Xe	131.30
Lithium	Li	6.939	Ytterbium	Yb	173.04
Lutetium	Lu	174.97	Yttrium	Y	88.905
Magnesium	Mg	24.312	Zinc	Zn	65.37
Manganese	Mn	54.9380	Zirconium	Zr	91.22
Mendelevium	Md	(256)			

MATHEMATICS

VOLUME AND SURFACE AREAS

A = cross sectional area or end area
S = lateral surface area
V = enclosed volume

RIGHT CIRCULAR CYLINDER

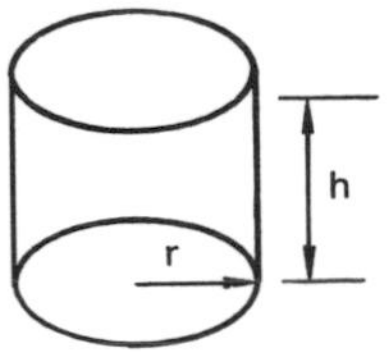

$A = \pi r^2$
$S = 2\pi rh$
$V = Ah = \pi r^2 h$

SPHERE

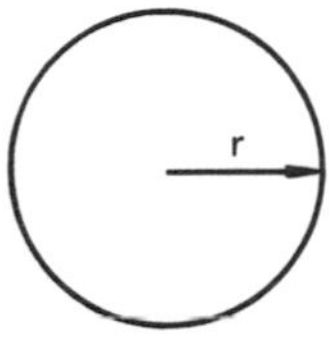

$S = 4\pi r^2$
$V = \left(\frac{4}{3}\right)\pi r^3$

RIGHT CIRCULAR CONE

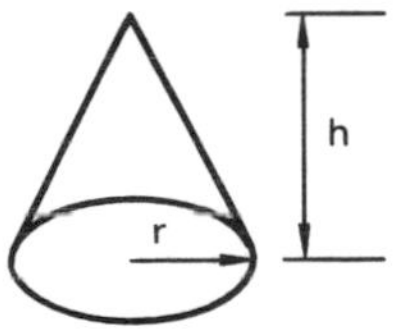

$A_{base} = \pi r^2$
$S = \pi r\sqrt{r^2 + h^2}$
$V = \left(\frac{1}{3}\right)\pi r^2 h$

MENSURATION

CIRCLE of radius r:

$$\text{circumference} = 2\pi r = p$$

$$\text{area} = \pi r^2 = \frac{p^2}{4\pi}$$

ELLIPSE with axes a and b

$$\text{area} = \pi ab$$

CIRCULAR SECTOR with radius r, included angle θ, and arc length s:

$$\text{area} = \tfrac{1}{2}\,\theta r^2 = \tfrac{1}{2}\,sr$$

$$\text{arc length} = s = \theta r$$

The angle θ is in radians.

QUADRATIC EQUATION

The roots of the equation $ax^2 + bx + c = 0$ are

$$\frac{-b \pm \sqrt{b^2 - 4ac}}{2a}$$

EXPONENTIATION

$$x^m x^n = x^{(n+m)}$$

$$\frac{x^m}{x^n} = x^{(m-n)}$$

$$(x^n)^m = x^{(mn)}$$

$$a^{m/n} = \sqrt[n]{a^m}$$

$$\left(\frac{a}{b}\right)^n = \frac{a^n}{b^n}$$

$$\sqrt[n]{x} = (x)^{1/n}$$

$$x^{-n} = \frac{1}{x^n}$$

$$x^0 = 1$$

LOGARITHM IDENTITIES

$$x^a = \text{antilog}\,[a \log (x)]$$

$$\log (x^a) = a \log (x)$$

$$\log (xy) = \log (x) + \log (y)$$

$$\log \left(\frac{x}{y}\right) = \log (x) - \log (y)$$

$$ln(x) = \frac{\log_{10} x}{\log_{10} e}$$

$$\approx 2.3\,(\log_{10} x)$$

$$\log_b(b) = 1$$

$$\log(1) = 0$$

$$\log_b(b^n) = n$$

TRIGONOMETRY

$\sin\theta = \frac{y}{h}$ $\cos\theta = \frac{x}{h}$ $\tan\theta = \frac{y}{x}$

$\csc\theta = \frac{h}{y}$ $\sec\theta = \frac{h}{x}$ $\cot\theta = \frac{x}{y}$

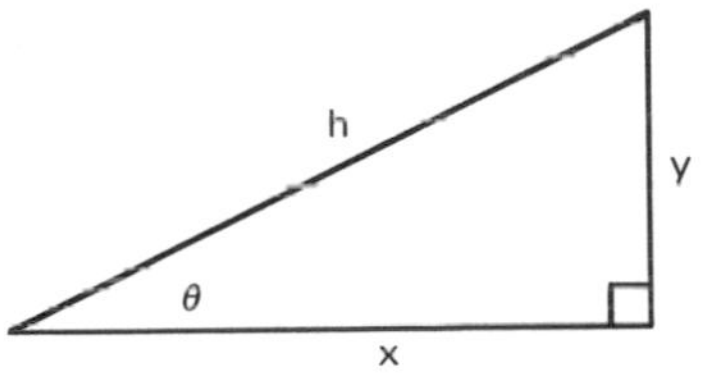

Basic Identities: $\sin^2\theta + \cos^2\theta = 1$

$$\sin 2\theta = 2\sin\theta\cos\theta$$

$$1 + \tan^2\theta = \sec^2\theta$$

$$\cos 2\theta = \cos^2\theta - \sin^2\theta = 2\cos^2\theta - 1$$

For general triangles: the law of sines is

$$\frac{\sin A}{a} = \frac{\sin B}{b} = \frac{\sin C}{c}$$

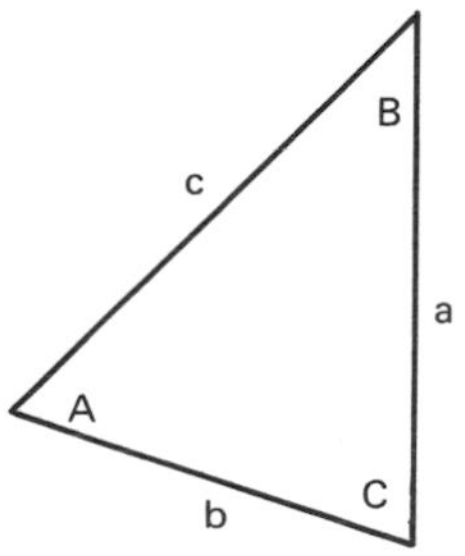

The law of cosines is

$$a^2 = b^2 + c^2 - 2bc\cos A$$

The area of a general triangle is ½ ab(sinC).

EQUATION OF A STRAIGHT LINE

The general form is $Ax + By + C = 0$.

The slope-intercept form is $y = mx + b$, where m is the slope and b is the y-intercept.

Given one point on the line (x^*,y^*), the point-slope form is

$$y - y^* = m(x - x^*).$$

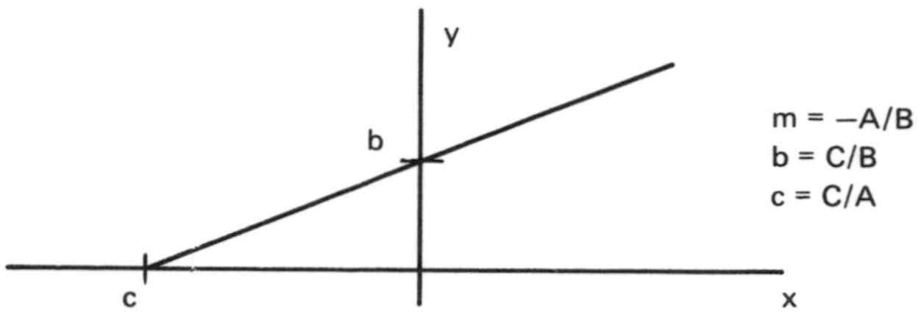

CONIC SECTIONS

The equation of a circle of radius r which is centered at (h,k) is

$$(x-h)^2 + (y-k)^2 = r^2.$$

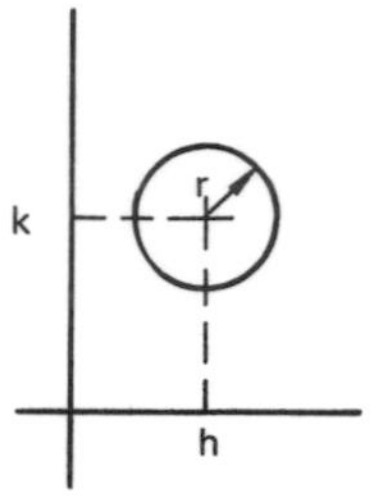

The equation of an ellipse centered at (h,k) with a and b defined as shown is

$$\left(\frac{x-h}{a}\right)^2 + \left(\frac{y-k}{b}\right)^2 = 1$$

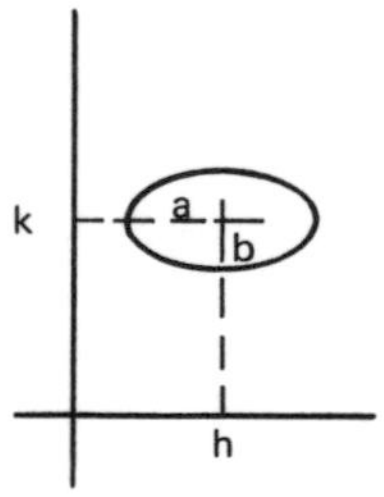

The equation of a parabola with vertex at (h,k) and symmetrical with respect to the y-axis is

$$(x-h)^2 = 4p\,(y-k)$$

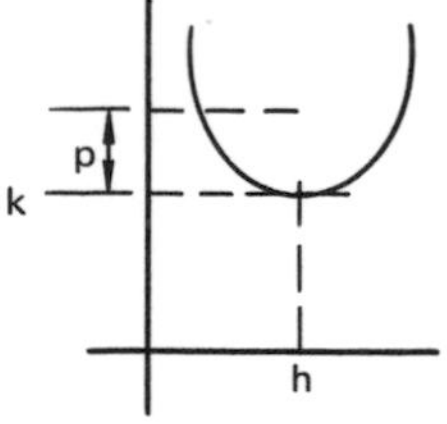

LINEAR REGRESSION

If it is necessary to draw a straight line through n data points $(x_1, y_1), (x_2, y_2), \ldots, (x_n, y_n)$, the following method based on the theory of least squares can be used:

step 1: Calculate the following quantities.

$$\Sigma x_i \quad \Sigma x_i^2 \quad (\Sigma x_i)^2 \quad \bar{x} = \left(\frac{\Sigma x_i}{n}\right) \quad \Sigma x_i y_i$$

$$\Sigma y_i \quad \Sigma y_i^2 \quad (\Sigma y_i)^2 \quad \bar{y} = \left(\frac{\Sigma y_i}{n}\right)$$

step 2: Calculate the slope of the line $y = mx + b$.

$$m = \frac{n\Sigma(x_i y_i) - (\Sigma x_i)(\Sigma y_i)}{n\Sigma x_i^2 - (\Sigma x_i)^2}$$

step 3: Calculate the y intercept.

$$b = \bar{y} - m\bar{x}$$

step 4: To determine the goodness of fit, calculate the correlation coefficient.

$$r = \frac{n\Sigma\,(x_i y_i) - (\Sigma x_i)(\Sigma y_i)}{\sqrt{[n\Sigma x_i^2 - (\Sigma x_i)^2]\,[n\Sigma y_i^2 - (\Sigma y_i)^2]}}$$

EXTREMA BY DIFFERENTIATION

Given a continuous function, f(x), the extreme points may be found by taking the first derivative and setting it equal to zero. Let x^* be the value of x which satisfies this equality. $f(x^*)$ is a minimum if $f''(x^*)$ is greater than zero. If $f''(x^*)$ is less than zero, $f(x^*)$ is a maximum. $f''(x^*)$ is equal to zero at an inflection point.

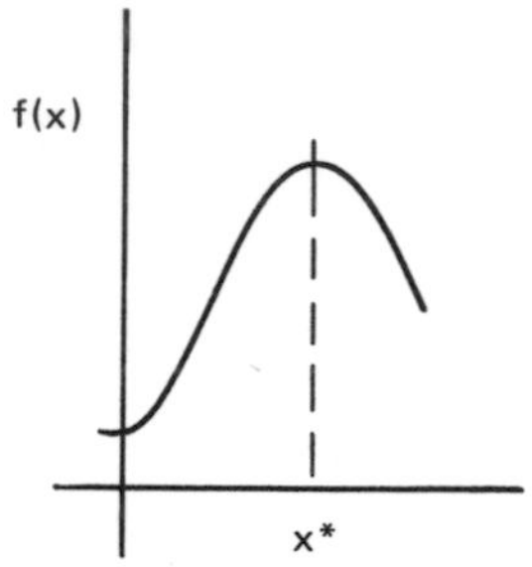

VECTORS

A vector is a directed line segment. It is defined completely only when both the magnitude and direction are known. Vectors are defined in terms of the unit vectors **i,j, k.** The unit vectors are vectors of length one directed along the $x,y,$ and z axes respectively. A vector, A, can be written in terms of the unit vectors and its endpoints (a_x, a_y, a_z).

$$A = a_x\mathbf{i} + a_y\mathbf{j} + a_z\mathbf{k}$$

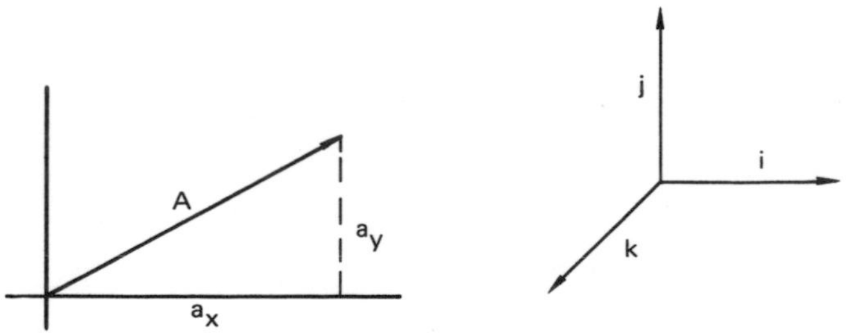

Addition of vectors is performed by adding the components:

$$A+B = (a_x+b_x)\mathbf{i} + (a_y+b_y)\mathbf{j} + (a_z+b_z)\mathbf{k}$$

The DOT PRODUCT is a scalar and represents the projection of B onto A.

$$A\cdot B = a_x b_x + a_y b_y = |A||B|\cos\theta$$

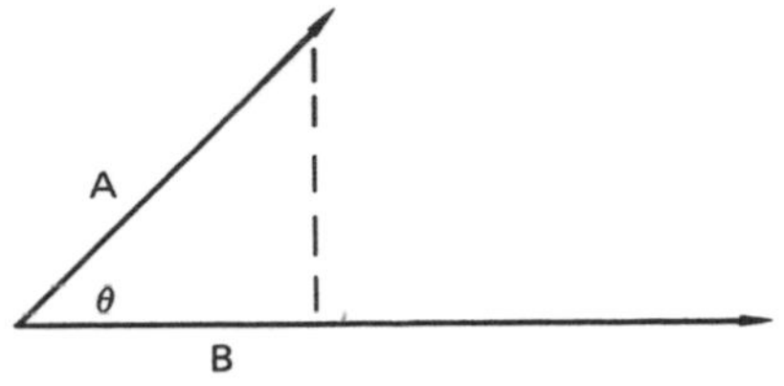

The CROSS PRODUCT is a vector of magnitude $|B||A|\sin\theta$ which is perpendicular to the plane containing A and B.

$$A \times B = \begin{vmatrix} \mathbf{i} & \mathbf{j} & \mathbf{k} \\ a_x & a_y & a_z \\ b_x & a_y & a_z \end{vmatrix}$$

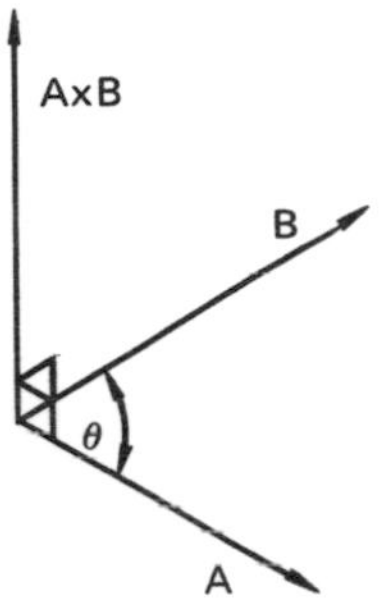

The sense of $A \times B$ is determined by the righthand rule.

BINOMIAL PROBABILITY DISTRIBUTION

f(x) is the probability that *x* will occur in *n* trials. *p* is the probability of one success in one trial. $q = (1-p)$ is the probability of one failure in one trial.

$$f(x) = \binom{n}{x} p^x q^{n-x} = \frac{n!}{(n-x)!x!} p^x q^{n-x}$$

The mean is *np*. The variance is *npq*.

POISSON PROBABILITY DISTRIBUTION

f(x) is the probability of *x* occurrences. *x* is the actual number of occurrences in some period. λ is the mean number of occurrences per period.

$$f(x) = \frac{e^{-\lambda}\lambda^x}{x!}$$

The mean and variance are both λ.

STATISTICS

$$\text{arithmetic mean} = \bar{x} = \left(\frac{1}{n}\right)(x_1 + x_2 + \ldots x_n) = \frac{\Sigma x_i}{n}$$

$$\text{geometric mean} = \sqrt[n]{x_1 x_2 x_3 \ldots x_n}$$

$$\text{harmonic mean} = \frac{n}{\frac{1}{x_1} + \frac{1}{x_2} + \ldots + \frac{1}{x_n}}$$

$$\text{root-mean-squared value} = \sqrt{\frac{\Sigma x_i^2}{n}}$$

$$\text{standard deviation} = \sigma = \sqrt{\frac{\Sigma(x_i - \bar{x})^2}{n}} = \sqrt{\frac{\Sigma x_i^2}{n} - (\bar{x})^2}$$

$$\text{sample standard deviation} = s = \sqrt{\frac{\Sigma(x_i - \bar{x})^2}{n-1}} = \sqrt{\frac{\Sigma x_i^2 - \frac{(\Sigma x_i)^2}{n}}{n-1}}$$

population variance = σ^2

sample variance = s^2

ENGINEERING ECONOMICS

NOMENCLATURE AND DEFINITIONS

A	annual amount or annuity: a payment or receipt of a fixed sum of money at yearly intervals.
C	cost: the asset purchase price.
d	declining balance depreciation rate: For double-declining balance depreciation, $d = \left(\frac{2}{n}\right)$.
D_j	depreciation in year j.
F	future worth, value, or amount: the value of an asset at some future point in time.
G	uniform gradient amount: a quantity by which the cash flows change each year.
i	annual effective interest rate.
k	number of compounding periods per year.
MARR	Minimum Attractive Rate of Return. Usually the same as the annual effective interest rate, *i*.
n	number of compounding periods; or the expected life of an asset.
P	present worth, value, or amount. The value of an amount at time = 0.
r	nominal annual interest rate. Same as rate per annum.
S_n	expected salvage value in year *n*.
t	time or income tax rate.
ϕ	effective interest rate per period. Equal to $\left(\frac{r}{k}\right)$.

YEAR-END CONVENTION

All cash disbursements and receipts (cash flows) are assumed to occur at the end of the year in which they actually occur. The exception is an initial cost (purchase price) which is assumed to occur at time = 0.

SUNK COSTS

Sunk costs are expenses incurred before time = 0. They have no bearing on the evaluation of alternatives.

CASH FLOW DIAGRAMS

Cash flow diagrams are drawn to help visualize problems involving transfers of money at various points in time. The following conventions are used in their construction:

- The horizontal axis represents time. The axis is divided into equal increments, one per period.
- The year-end convention is assumed.
- Disbursements are downward arrows; receipts are upward arrows. In both cases, the length of the arrow is proportional to the magnitude of the transfer.
- Two or more transfers in the same time period are represented by end-to-end arrows.

DISCOUNTING FACTORS

Discounting factors are numbers used to calculate the equivalent amount (at some point in time) of an alternative. The factors are given the functional notation (X/Y,i%,n) where *X* is the desired value, *Y* is the known value, *i* is the interest rate, and *n* is the number of periods. (Usually, *n* is in years.)

Single Payment Compound Amount	(F/P,i%,n)	$(1+i)^n$
Present Worth	(P/F,i%,n)	$(1+i)^{-n}$
Uniform Series Sinking Fund	(A/F,i%,n)	$\frac{i}{[(1+i)^n-1]}$
Capital Recovery	(A/P,i%,n)	$\frac{i(1+i)^n}{[(1+i)^n-1]}$
Compound Amount	(F/A,i%,n)	$\frac{[(1+i)^n-1]}{i}$
Equal Series Present Worth	(P/A,i%,n)	$\frac{[(1+i)^n-1)}{i(1+i)^n}$
Uniform Gradient	(P/G,i%,n)	$\left[\frac{1}{i} - \frac{n}{(1+i)^n - 1}\right]$ (P/A,i%,n)

COMPARING ALTERNATIVES

The PRESENT WORTH method may be used if all of the alternatives have equal lives. The present worth of each alternative is calculated. The alternative with the smallest negative or largest positive present worth is chosen.

The EQUIVALENT UNIFORM ANNUAL COST (EUAC) method must be used if alternatives have unequal lives. Use of this method is restricted to alternatives which are infinitely renewed. Specifically, each alternative is replaced by an identical replacement at the end of its useful life, up to the duration of the longest-lived alternative. The EUAC of an alternative is usually found from the following formula:

$$\text{EUAC} = (\text{present worth of alternative})\,(A/P,i\%,n)$$

The CAPITALIZED COST of a project is the present worth of a project which has an infinite life. The capitalized cost represents the amount of money needed at time = 0 to support the project on interest only, without reducing the principal.

$$\text{capitalized cost} = \text{initial cost} + \frac{(\text{annual maintenance cost})}{i}$$

TREATMENT OF SALVAGE VALUE IN REPLACEMENT STUDIES

By convention, the salvage value is subtracted from the defender's present value. This is done to keep all costs and benefits related to the defender with the defender. In this case, the salvage value is treated as an opportunity cost which would be incurred if the defender is not retired.

RATE OF RETURN

The rate of return (ROR) is the interest rate which makes the present worth equal to zero. To determine the ROR, assume a reasonable value for the interest rate earned and find the present worth. If the present worth is zero, the assumed interest rate is the ROR. If the present worth is not zero, assume another value of *i* and find the present worth. Use the two values of *i* and their corresponding present worths to interpolate or extrapolate a value of *i* which makes the present worth zero.

NON-ANNUAL COMPOUNDING

For problems in which compounding is done at intervals other than yearly, an effective annual interest rate can be computed.

$$i = \left[1 + \left(\frac{r}{k}\right)\right]^k - 1 = (1 + \phi)^k - 1$$

DEPRECIATION

STRAIGHT LINE	$D_j = \frac{(C - S_n)}{n}$
DOUBLE-DECLINING BALANCE (i = 1 to j−1)	$D_j = \frac{2(C - \Sigma D_i)}{n}$
SUM OF THE YEAR'S DIGITS	$D_j = \frac{(C - S_n)(n - j + 1)}{T}$
	$T = \frac{1}{2}n(n + 1)$

SINKING FUND $D_j = (C - S_n)(A/F,i\%,n)\ (F/P,i,n-1)$

ACRS $D_j = (\text{factor})\ C$

INCOME TAXES

If income taxes are paid, operating expenses and depreciation are deductible. If t is the tax rate, revenues and all expenses (except depreciation) should be multiplied by (1-t) in the year in which they occur. Although depreciation is a deductible expense, it is not an actual out-of-pocket expense. Depreciation should be multiplied by t and added to the cash flow in the appropriate year.

CONSUMER LOANS

BAL_j	balance after the *j*th payment
LV	total value loaned (cost minus down payment)
j	payment or period number
N	total number of payments to pay off the loan
PI_j	*j*th interest payment
PP_j	*j*th principal payment
PT_j	*j*th total payment
ϕ	effective rate per period $\left(\frac{r}{k}\right)$

SIMPLE INTEREST

Interest due does not compound with a *simple interest* loan. The interest due is proportional to the length of time the principal is outstanding.

DIRECT REDUCTION LOANS

This is the typical 'interest paid on unpaid balance' loan. The amount of the periodic payment is constant, but the amounts paid towards the principal and interest both vary.

$$N = -\frac{ln\left(\frac{-\phi(LV)}{PT} + 1\right)}{ln(1 + \phi)}$$

$$PT = LV\,(A/P,\ i\%,\ n)$$

$$LV = \frac{PT}{(A/P,\ i\%,\ n)}$$

$$BAL_{j-1} = PT\left[\frac{1 - (1 + \phi)^{j-1-N}}{\phi}\right]$$

$$PI_j = \phi(BAL_j)$$

$$PP_j = PT - PI_j$$

$$BAL_j = BAL_{j-1} - PP_j$$

HANDLING INFLATION

It is important to perform economic studies in terms of *constant value dollars*. One method of converting all cash flows to constant value dollars is to divide the flows by some annual *economic indicator* or price index.

An alternative is to replace i with a value corrected for the inflation rate, e. This corrected value, i', is

$$i' = i + e + ie$$

FLUID STATICS AND DYNAMICS

IMPORTANT FLUID CONVERSIONS

multiply	by	to get
cubic feet	7.4805	gallons
cfs	448.83	gpm
cfs	.64632	MGD
gallons	.1337	cubic feet
gpm	.002228	cfs
inches of mercury	.491	psi
inches of mercury	70.7	psf
inches of mercury	13.60	inches of water
inches of water	5.199	psf
inches of water	.0361	psi
inches of water	.0735	inches of mercury
psi	144	psf
psi	2.308	feet of water
psi	27.7	inches of water
psi	2.037	inches of mercury
psf	.006944	psi

DENSITY

density of water: 62.4 lbm/ft^3, .0361 lbm/in^3

density of mercury: 848.4 lbm/ft^3, .491 lbm/in^3

VISCOSITY

μ: absolute viscosity $\left(\frac{lbf\text{-}sec}{ft^2}\right)$

ν: kinematic viscosity $\left(\frac{ft^2}{sec}\right) = \frac{\mu g}{\rho}$

TYPICAL VISCOSITY UNITS

	Absolute	Kinematic
English	lbf-sec/ft² (slug/ft-sec)	ft²/sec
Conventional Metric	dyne-sec/cm² (poise)	cm²/sec (stoke)
SI	Pascal-second (N-s/m²)	m²/s

HYDROSTATIC PRESSURE

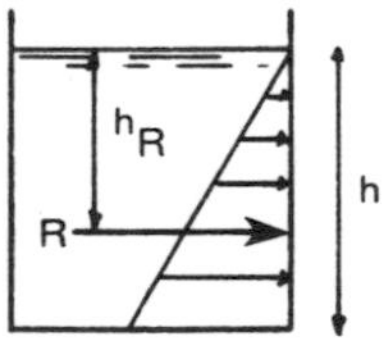

$$p = \rho h$$
$$\bar{p} = \tfrac{1}{2}\rho h$$
$$R = \bar{p}A$$
$$h_R = \left(\tfrac{2}{3}\right) h$$

PRESSURE ON SUBMERGED PLANE SURFACES

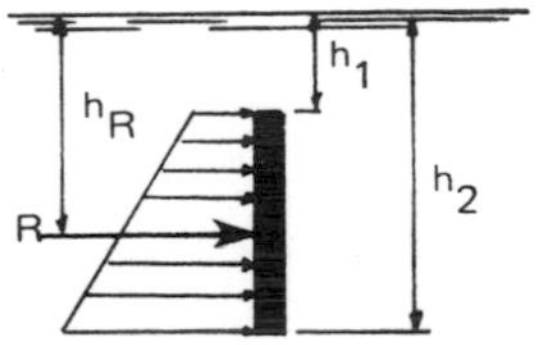

h_c = distance from surface to centroid of plane, as measured parallel to plane's surface

$$\bar{p} = \rho h_c$$
$$R = \bar{p}A$$
$$h_R = h_c + \left(\frac{I}{Ah_c}\right) \text{ as measured parallel to plane's surface}$$

MANOMETERS

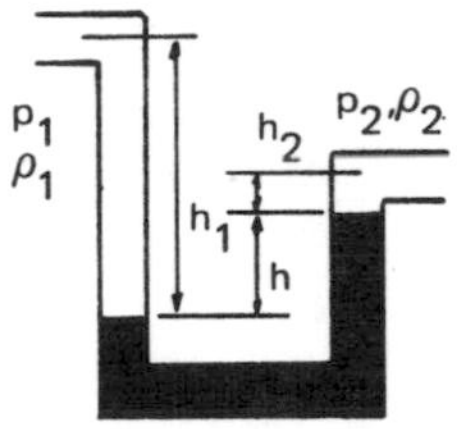

$$\Delta p = p_1 - p_2$$
$$\Delta p = \rho_m h - \rho_1 h_1 + \rho_2 h_2$$

BUOYANCY (ARCHIMEDES' PRINCIPLE)

The buoyant force is equal to the weight of the displaced fluid.

SPEED OF SOUND IN A FLUID

$$c = \sqrt{kg_cRT} \quad \text{(gases)}$$
$$c = \sqrt{\frac{Eg_c}{\rho}} \quad \text{(liquids)}$$

E (bulk modulus) $\approx$ 4.3 EE7 lbf/ft^2 for water

CONTINUITY EQUATION

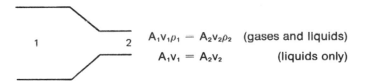

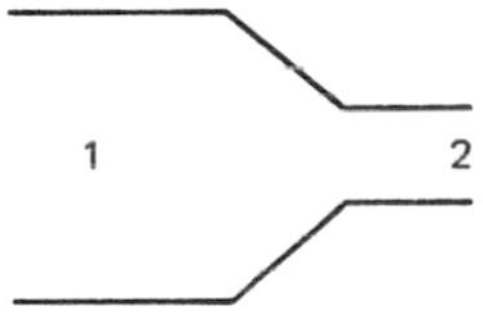

$$A_1v_1\rho_1 = A_2v_2\rho_2 \quad \text{(gases and liquids)}$$
$$A_1v_1 = A_2v_2 \quad \text{(liquids only)}$$

HYDRAULIC RADIUS

The hydraulic radius is defined as the area in flow divided by the wetted perimeter. The wetted perimeter does not include free fluid surface.

$$r_h = \frac{\text{area in flow}}{\text{wetted perimeter}} = D_e/4$$

BERNOULLI EQUATION

$$\frac{p}{\rho} + \frac{v^2}{2g_c} + z = \text{total head}$$

$\left(\frac{p}{\rho}\right)$ = pressure (static) head

$\left(\frac{v^2}{2g_c}\right)$ = velocity (dynamic) head

z = potential (gravitational) head

CONSERVATION OF ENERGY

$$\frac{p_1}{\rho} + \frac{v_1^2}{2g_c} + z_1 + w_{pump} = \frac{p_2}{\rho} + \frac{v_2^2}{2g_c} + z_2 + h_f + w_{turbine}$$

w_{pump} and $w_{turbine}$ are on a per-pound basis with units of (ft-lbf/lbm).

REYNOLDS NUMBER

$$N_{Re} = \frac{vD\rho}{\mu g_c} = \frac{vD}{\nu}$$

DARCY FRICTION LOSS AND MINOR LOSS

$$h_f = \frac{fLv^2}{2Dg_c} = \frac{Kv^2}{2g_c}$$

K is the friction loss coefficient.

PUMP TERMS

$$W_{pump} = \dot{m}h \text{ in ft-lbf/sec}$$

$$\dot{m} = (\text{cfs})(\rho) \text{ in lbm/sec}$$

$$h = \text{added head in feet}$$

$$hp = \frac{W_{pump}}{550\eta} = \frac{(\text{gpm})(h)(\text{S.G.})}{(3956)\eta}$$

HYDRAULIC GRADE LINE

The hydraulic grade line is a graphical representation of the sum of the static and potential heads versus position along the pipeline.

$$\text{hydraulic grade} = z + \frac{p}{\rho}$$

TORRICELLI'S EQUATION - DISCHARGE

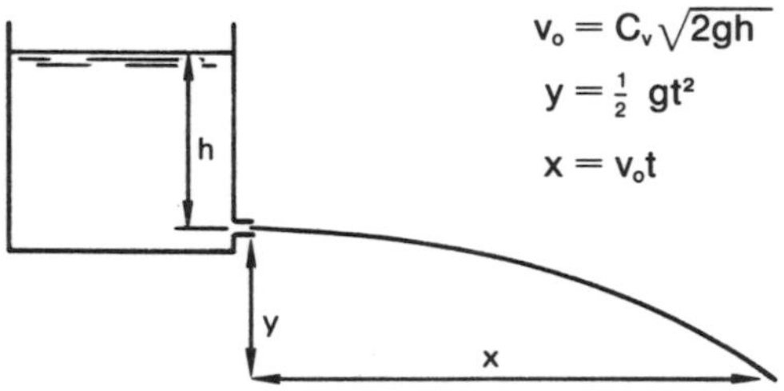

$$v_o = C_v\sqrt{2gh}$$

$$y = \tfrac{1}{2}gt^2$$

$$x = v_o t$$

For a tank with a constant cross-sectional area, the time required to lower the fluid level from level h_1 to h_2 is

$$t = \frac{2A_t(\sqrt{h_1} - \sqrt{h_2})}{C_d A_o \sqrt{2g}}$$

VENTURI METER

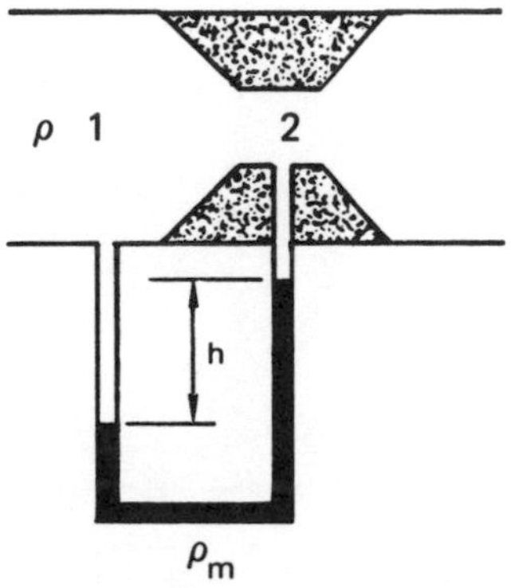

$$\beta = \left(\frac{D_2}{D_1}\right)$$

$$F_{va} = \frac{1}{\sqrt{1 - (\beta)^4}}$$

$$v_2 = F_{va}\sqrt{\frac{2gh(\rho_m - \rho)}{\rho}}$$

$$C_d = C_c C_v$$

$$C_c \approx 1 \text{ for venturi meters}$$

$$C_f = C_d F_{va}$$

$$Q = C_f A_2\sqrt{\frac{2gh(\rho_m - \rho)}{\rho}}$$

$$\rho \approx 0 \text{ for air}$$

ORIFICE PLATE

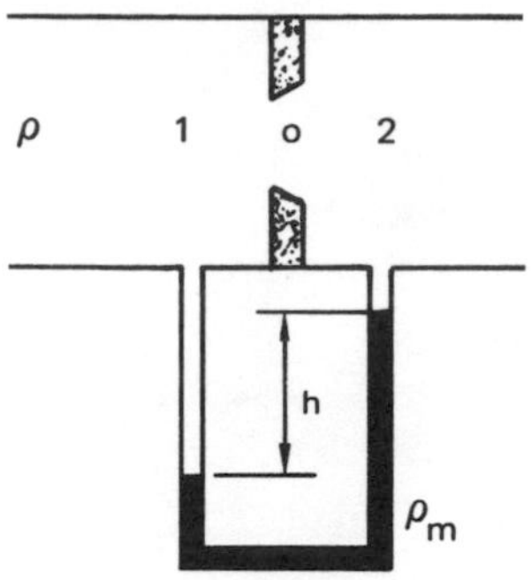

$$F_{va} = \frac{1}{\sqrt{1 - C_c^2\left(\frac{A_o}{A_1}\right)^2}}$$

$$v_2 = F_{va}C_v\sqrt{\frac{2gh(\rho_m - \rho)}{\rho}}$$

$$C_f = F_{va}C_d$$

$$C_d = C_c C_v$$

$$Q = C_c A_o v_2 = A_o C_f\sqrt{\frac{2gh(\rho_m - \rho)}{\rho}}$$

Q is in cfs

IMPULSE-MOMENTUM

Q is in cfs

$$\dot{m} = \frac{Q\rho}{g_c}$$

$$R_x = A_1 p_1 - A_2 p_2 \cos\phi - \dot{m}(v_2\cos\phi - v_1)$$

$$R_y = A_2 p_2 \sin\phi + \dot{m}v_2\sin\phi$$

p's are gage pressures

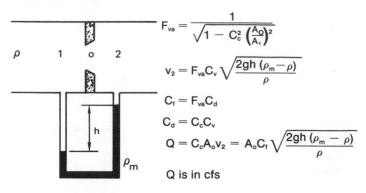

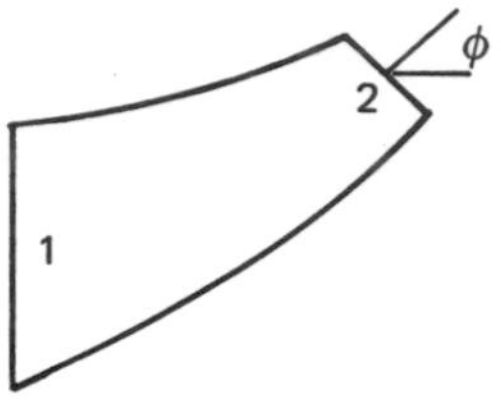

LIFT AND DRAG

$$L = \frac{C_L v^2 \rho A}{2g_c}$$

$$D = \frac{C_D v^2 \rho A}{2g_c}$$

A is the projected area (circle for a sphere).

SIMILARITY

For fans, pumps, turbines, drainage through holes in tanks, closed-pipe flow with no free surfaces (in the turbulent region with the same relative roughness), and for completely submerged objects such as torpedoes, airfoils, and submarines, performance of models and prototypes can be correlated by equating Reynolds numbers.

$$(N_{Re})_{model} = (N_{Re})_{prototype}$$

HYDRAULIC MACHINES

NET POSITIVE SUCTION HEAD

$$NPSHA = h_{atmos} + h_{static} - h_{friction} - h_{vapor}$$
$$= h_{pressure\,(i)} + h_{velocity\,(i)} - h_{vapor}$$
$$\frac{NPSHR_2}{NPSHR_1} = \left(\frac{Q_2}{Q_1}\right)^2$$

PUMPING POWER (HEAD ADDED)

$$h_A = \frac{p_d}{\rho} - \frac{p_i}{\rho} + \frac{v_d^2}{2g_c} - \frac{v_i^2}{2g_c} + z_d - z_i$$

HYDRAULIC HORSEPOWER

	Q in gpm	$\dot{m}$ in lbm/sec	$\dot{V}$ in cfs
h_A is added head in feet	$\frac{h_A Q (S.G.)}{3956}$	$\frac{h_A \dot{m}}{550}$	$\frac{h_A \dot{V} (S.G.)}{8.814}$
p is added head in psf	$\frac{pQ}{2.468\ EE5}$	$\frac{p\dot{m}}{(34320)(S.G.)}$	$\frac{p\dot{V}}{550}$

SPECIFIC SPEED

$$n_s = \frac{n\sqrt{Q}}{(h_A)^{.75}} \qquad \text{(pumps)}$$

$$n_s = \frac{n\sqrt{bhp}}{(H)^{1.25}} \qquad \text{(turbines)}$$

CENTRIFUGAL PUMP IMPELLER TYPES

Approximate Range of Specific Speed (rpm)	Impeller Type
500–1000	radial vane
2000–3000	Francis (mixed) vane
4000–7000	mixed flow
9000 and above	axial flow

PUMP AFFINITY LAWS

$$\frac{Q_2}{Q_1} = \frac{n_2}{n_1}$$

$$\frac{h_2}{h_1} = \left(\frac{n_2}{n_1}\right)^2 = \left(\frac{Q_2}{Q_1}\right)^2$$

$$\frac{bhp_2}{bhp_1} = \left(\frac{n_2}{n_1}\right)^3 = \left(\frac{Q_2}{Q_1}\right)^3$$

$$\frac{Q_2}{Q_1} = \frac{d_2}{d_1}$$

$$\frac{h_2}{h_1} = \left(\frac{d_2}{d_1}\right)^2$$

$$\frac{bhp_2}{bhp_1} = \left(\frac{d_2}{d_1}\right)^3$$

PUMP SIMILARITY LAWS

$$\frac{n_1 d_1}{\sqrt{h_1}} = \frac{n_2 d_2}{\sqrt{h_2}}$$

$$\frac{Q_1}{d_1^2\sqrt{h_1}} = \frac{Q_2}{d_2^2\sqrt{h_2}}$$

$$\frac{bhp_1}{\rho_1 d_1^2 h_1^{1.5}} = \frac{bhp_2}{\rho_2 d_2^2 h_2^{1.5}}$$

$$\frac{Q_1}{n_1 d_1^3} = \frac{Q_2}{n_2 d_2^3}$$

$$\frac{bhp_1}{\rho_1 n_1^3 d_1^5} = \frac{bhp_2}{\rho_2 n_2^3 d_2^5}$$

$$\frac{n_1\sqrt{Q_1}}{(h_1)^{.75}} = \frac{n_2\sqrt{Q_2}}{(h_2)^{.75}}$$

FANS AND DUCTWORK

EQUATIONS AND CONVERSIONS

$$\text{inches w.g.} = \frac{\text{psig}}{.0361} \qquad \text{(static)}$$

$$= \left(\frac{V_{fpm}}{4005}\right)^2 \qquad \text{(dynamic)}$$

AIR HORSEPOWER

$$p_t = p_s + p_v \qquad \text{(inches w.g.)}$$

$$AHP = \frac{(\text{cfm})\, p_t}{6356}$$

SYSTEM LOSS CURVE

$$\frac{\Delta p_2}{\Delta p_1} = \left(\frac{Q_2}{Q_1}\right)^2$$

FAN LAWS

$$\frac{Q_A}{Q_B} = \left(\frac{D_A}{D_B}\right)^3 \left(\frac{n_A}{n_B}\right) = \left(\frac{D_A}{D_B}\right)^2 \sqrt{\frac{p_A}{p_B}} \sqrt{\frac{\rho_B}{\rho_A}}$$

$$\frac{p_A}{p_B} = \left(\frac{D_A}{D_B}\right)^2 \left(\frac{n_A}{n_B}\right)^2 \left(\frac{\rho_A}{\rho_B}\right)$$

$$\frac{AHP_A}{AHP_B} = \left(\frac{D_A}{D_B}\right)^5 \left(\frac{n_A}{n_B}\right)^3 \left(\frac{\rho_A}{\rho_B}\right)$$

$$= \left(\frac{D_A}{D_B}\right)^2 \left(\frac{p_A}{p_B}\right)^{1.5} \sqrt{\frac{\rho_B}{\rho_A}}$$

$$= \left(\frac{Q_A}{Q_B}\right) \left(\frac{p_A}{p_B}\right)$$

$$\frac{n_A}{n_B} = \left(\frac{D_B}{D_A}\right) \sqrt{\frac{p_A}{p_B}} \sqrt{\frac{\rho_B}{\rho_A}}$$

$$= \sqrt{\frac{Q_B}{Q_A}} \left(\frac{p_A}{p_B}\right)^{.75} \left(\frac{\rho_B}{\rho_A}\right)^{.75}$$

$$\frac{D_A}{D_B} = \sqrt{\frac{Q_A}{Q_B}} \left(\frac{p_B}{p_A}\right)^{.25} \left(\frac{\rho_A}{\rho_B}\right)^{.25}$$

EQUIVALENT DIAMETER OF RECTANGULAR DUCT

$$D_e = 1.3\, \frac{(ab)^{.625}}{(a+b)^{.25}}$$

MINOR DUCT LOSSES

$$\Delta p = c p_v = c \left(\frac{v}{4005}\right)^2$$

WINDSTREAM POWER CONTENT

$$P_{ideal} = \frac{\pi\, r_{rotor}^2\, \rho v^3}{2 g_c} \quad \left(\frac{\text{ft-lbf}}{\text{sec}}\right)$$

$$P_{actual} = C_P\, P_{ideal}$$

C_P = power coefficient

PROFESSIONAL PUBLICATIONS INC. • P.O. Box 199, San Carlos, CA 94070

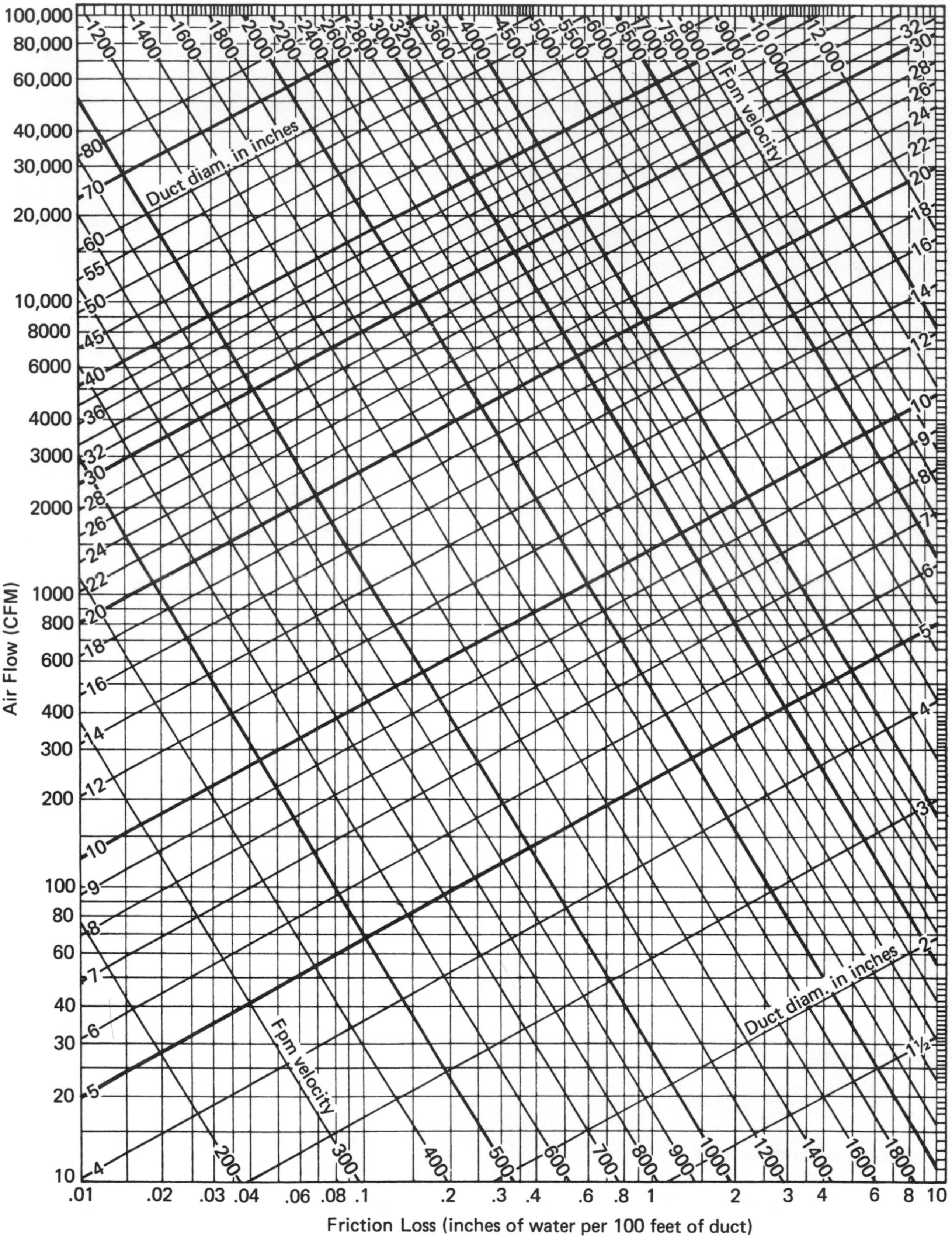

Air Flow (CFM)
Friction Loss (inches of water per 100 feet of duct)
Duct diam. in inches
Fpm velocity

THERMODYNAMICS

TYPES OF THERMODYNAMIC SYSTEMS

A system is a region with artificially-chosen boundaries. A closed system is one in which matter does not cross the system boundaries. Energy may cross the system boundaries, however. Both energy and matter may cross the boundaries for an open system.

MATHEMATICAL FORMULATION OF THE FIRST LAW

For closed systems:

$$dQ = dU + dW \quad \text{or} \quad Q = \Delta U + W$$

For steady-flow open systems:

$$Q = \frac{W}{J} + h_2 - h_1 + \frac{(z_2 - z_1)}{J} + \frac{(v_2^2 - v_1^2)}{2g_cJ}$$

(Q, W, and h are on a per-pound basis.)

TEMPERATURE SCALES

$$°R = °F + 460$$
$$°K = °C + 273$$
$$°F = 32 + (9/5)\,°C$$
$$°C = (5/9)(°F - 32)$$
$$\Delta °R = (9/5)\Delta °K$$
$$\Delta °K = (5/9)\Delta °R$$

IMPORTANT CONVERSIONS

1 BTU = 778 ft-lbf
1 BTU = 252 calories
1 hp = 550 ft-lbf/sec
1 kw = 3413 BTU/hr
1 kw = 1.341 hp
1 kw-hr = 3413 BTU

STP (STANDARD TEMPERATURE AND PRESSURE)

Scientific STP is 32°F and 14.7 psia.
For fuel gases, STP is 60°F and 14.7 psia.

COMPOSITION OF DRY ATMOSPHERIC AIR

(Rare inert gases included as N_2)

	% by weight	% by volume
Oxygen (O_2)	23.15	20.9
Nitrogen (N_2)	76.85	79.1

IDEAL GAS EQUATION OF STATE

$pV = nR^*T$ (for n moles)
$pV = mRT$ (for m pounds)
$R^* = 1545$ ft-lbf/pmole-°R
$R = 53.3$ ft/°R for air

COMPRESSED GAS EQUATION OF STATE

$pV = mZRT$
Z = compressibility factor

IDEAL GAS PROCESS LAW

$$\frac{p_1V_1}{T_1} = \frac{p_2V_2}{T_2}$$

SPECIFIC HEAT RELATIONSHIPS FOR IDEAL GASES

$c_p = 0.24$ BTU/lbm-°F (for air)
$c_v = 0.1714$ BTU/lbm-°F (for air)
$k = c_p/c_v = 1.4$ (for air)
$c_p - c_v = (R/J)$
$c_p = Rk/J(k-1)$

MIXTURES OF IDEAL GASES

volumetrically weighted: density and molecular weight of the mixture

gravimetrically weighted: internal energy, enthalpy, entropy, specific heats, and specific gas constant of the mixture

THERMODYNAMIC RELATIONSHIPS FOR ANY PROCESS (FOR IDEAL GASES)

$$\Delta u = c_v \Delta T$$
$$\Delta h = c_p \Delta T$$

TYPES OF PROCESSES

Isochoric (isometric) — constant volume
Isobaric — constant pressure
Isothermal — constant temperature
Polytropic — any process for which $p(V)^n$ is constant
Adiabatic — a process with no heat transfer
  Isentropic — an adiabatic process for which $\Delta s = 0$
  Throttling — an adiabatic process for which $\Delta h = 0$

SENSIBLE HEAT

$$Q = mc(T_2 - T_1)$$

$$c = \frac{1 \text{ BTU}}{\text{lbm}-°F} = \frac{1 \text{ cal}}{\text{gram}-°C} \quad \text{(for water)}$$

LATENT HEAT

Latent heat is the energy which changes the phase of a substance.
Latent heat of fusion: 143.4 BTU/lbm for turning ice to liquid
Latent heat of vaporization: 970.3 BTU/lbm for turning liquid to steam
Latent heat of sublimation: 1293.5 BTU/lbm for turning ice to steam directly
The above values are for water phases at 1 atmosphere.

QUALITY OF A LIQUID-VAPOR MIXTURE

$$x = \frac{\text{weight of vapor}}{\text{weight of liquid} + \text{weight of vapor}}$$

PROPERTIES OF A LIQUID-VAPOR MIXTURE

$$h = h_F + xh_{FG}$$
$$u = u_F + xu_{FG}$$
$$\upsilon = \upsilon_F + x\upsilon_{FG}$$
$$s = s_F + xs_{FG}$$

CONSTANT PRESSURE, CLOSED SYSTEMS

$$p_2 = p_1$$
$$T_2 = T_1\left(\frac{\upsilon_2}{\upsilon_1}\right)$$
$$\upsilon_2 = \upsilon_1\left(\frac{T_2}{T_1}\right)$$
$$Q = h_2 - h_1$$
$$= c_p(T_2 - T_1)$$
$$= c_v(T_2 - T_1) + p(\upsilon_2 - \upsilon_1)$$

$$u_2 - u_1 = c_v(T_2 - T_1)$$
$$= \frac{c_v p(v_2 - v_1)}{R}$$
$$= \frac{p(v_2 - v_1)}{k - 1}$$
$$W = p(v_2 - v_1)$$
$$= R(T_2 - T_1)$$
$$s_2 - s_1 = c_p ln\left(\frac{T_2}{T_1}\right)$$
$$= c_p ln\left(\frac{v_2}{v_1}\right)$$
$$h_2 - h_1 = Q$$
$$= \frac{kp(v_2 - v_1)}{k - 1}$$

CONSTANT VOLUME, CLOSED SYSTEMS

$$p_2 = p_1\left(\frac{T_2}{T_1}\right)$$
$$T_2 = T_1\left(\frac{p_2}{p_1}\right)$$
$$v_2 = v_1$$
$$Q = u_2 - u_1$$
$$= c_v(T_2 - T_1)$$
$$u_2 - u_1 = Q$$
$$= \frac{c_v v(p_2 - p_1)}{R}$$
$$= \frac{v(p_2 - p_1)}{k - 1}$$
$$W = 0$$
$$s_2 - s_1 = c_v ln\left(\frac{T_2}{T_1}\right)$$
$$= c_v ln\left(\frac{p_2}{p_1}\right)$$
$$h_2 - h_1 = c_p(T_2 - T_1)$$
$$= \frac{kv(p_2 - p_1)}{k - 1}$$

CONSTANT TEMPERATURE, CLOSED SYSTEMS

$$p_2 = p_1\left(\frac{v_1}{v_2}\right)$$
$$T_2 = T_1$$
$$v_2 = v_1\left(\frac{p_1}{p_2}\right)$$
$$Q = W$$
$$= T(s_2 - s_1)$$
$$= p_1 v_1 ln\left(\frac{v_2}{v_1}\right)$$
$$= RT ln\left(\frac{v_2}{v_1}\right)$$
$$u_2 - u_1 = 0$$
$$W = Q$$
$$= RT ln\left(\frac{p_1}{p_2}\right)$$
$$s_2 - s_1 = \frac{Q}{T}$$
$$= R ln\left(\frac{p_1}{p_2}\right)$$
$$h_2 - h_1 = 0$$

POLYTROPIC, CLOSED SYSTEMS

To use with isentropic, closed systems, substitute k for n in all equations.

$$p_2 = p_1\left(\frac{v_1}{v_2}\right)^n$$
$$= p_1\left(\frac{T_2}{T_1}\right)^{\frac{n}{n-1}}$$
$$T_2 = T_1\left(\frac{v_1}{v_2}\right)^{n-1}$$
$$= T_1\left(\frac{p_2}{p_1}\right)^{\frac{n-1}{n}}$$
$$v_2 = v_1\left(\frac{p_1}{p_2}\right)^{\frac{1}{n}}$$
$$= v_1\left(\frac{T_1}{T_2}\right)^{\frac{1}{n-1}}$$
$$Q = \frac{c_v(n - k)(T_2 - T_1)}{n - 1}$$
$$u_2 - u_1 = c_v(T_2 - T_1)$$
$$= \frac{p_2 v_2 - p_1 v_1}{n - 1}$$
$$W = \frac{R(T_1 - T_2)}{n - 1}$$
$$= \frac{p_1 v_1 - p_2 v_2}{n - 1}$$
$$= \frac{p_1 v_1}{n - 1}\left[1 - \left(\frac{p_2}{p_1}\right)^{\frac{n-1}{n}}\right]$$
$$s_2 - s_1 = \frac{c_v(n - k)}{n - 1}\left[ln\left(\frac{T_2}{T_1}\right)\right]$$
$$h_2 - h_1 = c_p(T_2 - T_1)$$
$$= \frac{n(p_2 v_2 - p_1 v_1)}{n - 1}$$

ISENTROPIC, STEADY FLOW SYSTEMS

p_2, v_2, and T_2 are the same as for isentropic, closed systems.

$$Q = 0$$
$$W = h_1 - h_2$$
$$= c_p T_1\left[1 - \left(\frac{p_2}{p_1}\right)^{\frac{k-1}{k}}\right]$$
$$u_2 - u_1 = c_v(T_2 - T_1)$$
$$h_1 - h_2 = W$$
$$= c_p(T_2 - T_1)$$
$$= \frac{k(p_2 v_2 - p_1 v_1)}{k - 1}$$
$$s_2 - s_1 = 0$$

VAPOR, COMBUSTION, AND REFRIGERATION CYCLES

ENERGY, WORK, AND POWER CONVERSIONS

multiply	by	to get
BTU	3.929 EE−4	hp-hrs
BTU	778.3	ft-lbf
BTU	2.930 EE−4	kw-hrs
BTU	1.0 EE−5	therms
BTU/hr	.2161	ft-lbf/sec
BTU/hr	3.929 EE−4	hp
BTU/hr	.2930	watts
ft-lbf	1.285 EE−3	BTU
ft-lbf	3.766 EE−7	kw-hrs
ft-lbf	5.051 EE−7	hp-hrs
ft-lbf/sec	4.6272	BTU/hr
ft-lbf/sec	1.818 EE−3	hp
ft-lbf/sec	1.356 EE−3	kw
hp	2545.0	BTU/hr
hp	550	ft-lbf/sec
hp	.7457	kw
hp-hr	2545.0	BTU
hp-hr	1.976 EE6	ft-lbf
hp-hr	.7457	kw-hrs
kw	1.341	hp
kw	3412.9	BTU/hr
kw	737.6	ft-lbf/sec

ENERGY AND POWER CONVERSIONS

1 BTU = 778 ft-lbf
1 hp = 550 ft-lbf/sec
1 kw = 3413 BTU/hr = 1.341 hp
1 therm = 100,000 BTU
1 kw-hr = 3413 BTU

GENERAL POWER CYCLE

The general power cycle moves clockwise on the p-v and T-s diagrams.

1→2: compression
2→3: heat addition
3→4: expansion
4→1: heat rejection

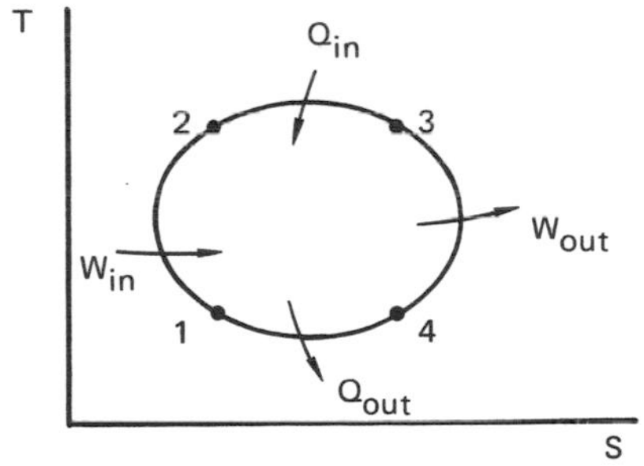

THERMAL EFFICIENCY OF THE ENTIRE CYCLE

$$\eta_{th} = \frac{Q_{in} - Q_{out}}{Q_{in}} = \frac{W_{out} - W_{in}}{Q_{in}}$$

CARNOT CYCLE EFFICIENCY

For the Carnot cycle, the thermal efficiency does not depend on the working fluid. It can be calculated directly from the two temperature extremes.

$$\eta_{th,\ Carnot} = \frac{T_{high} - T_{low}}{T_{high}}$$

ISENTROPIC EFFICIENCY OF A PUMP

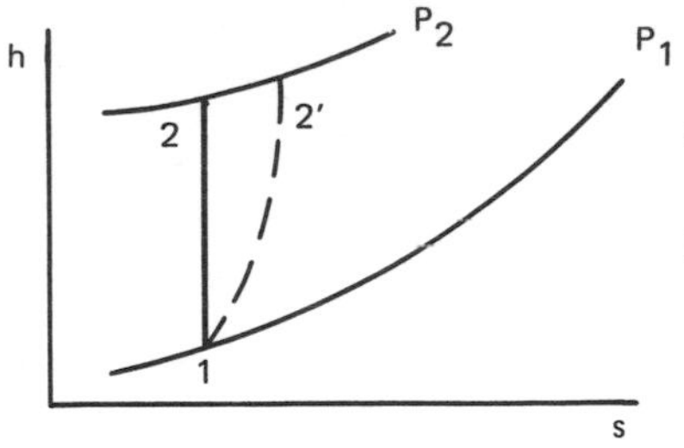

$$\eta = \frac{h_2 - h_1}{h_2' - h_1}$$

$$h_2' = h_1 + \frac{(h_2 - h_1)}{\eta}$$

ISENTROPIC EFFICIENCY OF A TURBINE

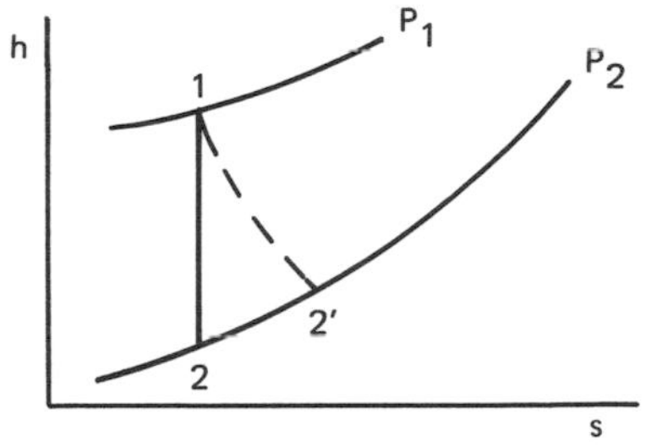

$$\eta = \frac{h_1 - h_2'}{h_1 - h_2}$$

$$h_2' = h_1 - \eta\,(h_1 - h_2)$$

RANKINE CYCLE WITH SUPERHEATING

$$W_{turb} = h_d - h_e$$

$$W_{pump} = v_f\,(p_a - p_f) = h_a - h_f$$

$$Q_{in} = h_d - h_a$$

$$Q_{out} = h_e - h_f$$

The thermal efficiency of the entire cycle is

$$\eta_{th} = \frac{Q_{in} - Q_{out}}{Q_{in}} = \frac{W_{turb} - W_{pump}}{Q_{in}}$$

$$= \frac{(h_d - h_a) - (h_e - h_f)}{h_d - h_a}$$

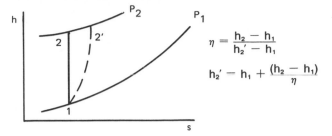

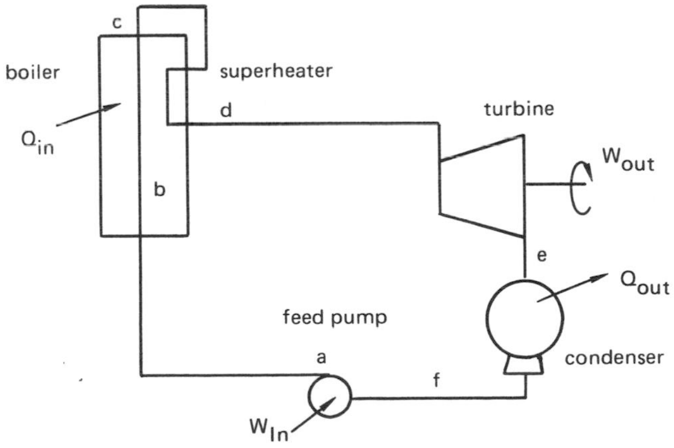

If the Rankine cycle gives efficiencies for the turbine and the pump, calculate all quantities as if those efficiencies were 100%. Then modify h_e and h_a prior to recalculating the thermal efficiency.

$$h_e' = h_d - \eta_{turb}\,(h_d - h_e)$$

$$h_a' = h_f + \frac{(h_a - h_f)}{\eta_{pump}}$$

$$W'_{turb} = h_d - h'_e$$

$$W'_{pump} = h'_a - h_f$$

$$Q'_{in} = h_d - h'_a$$

HEAT OF COMBUSTION

The heat of combustion of a fuel (also known as the 'heating value') is the amount of energy given off when a unit of fuel is burned. Units of heating value are BTU/lbm, BTU/gal, and BTU/ft^3 depending on whether the fuel is solid, liquid, or gas. The higher heating value (HHV) includes the heat of vaporization of the water vapor formed.

RATE OF FUEL CONSUMPTION

The rate of fuel consumption in internal combustion engines is known as the 'specific fuel consumption' (SFC) with units of lbm/hp-hr.

$$\text{hourly fuel consumption} = (\text{hp})(\text{SFC})$$

THE PLAN FORMULA

$$\text{hp} = \frac{\text{PLAN}}{33{,}000}$$

$$N = \frac{(2n)(\text{no. cylinders})}{(\text{no. strokes/cycle})}$$

P is the mean effective pressure in psig
L is the stroke length in feet
A is the bore area in in^2
N is the number of power strokes per minute
n is the engine speed in rpm

HORSEPOWER VERSUS TORQUE

$$(\text{hp})(5252) = (T \text{ in ft-lbf})(\text{rpm})$$

FLOW THROUGH NOZZLES

$$v = \sqrt{2g_c(h_1 - h_2)}$$

GENERAL REFRIGERATION CYCLE

The general refrigeration cycle moves counter-clockwise on the p-v and T-s diagrams

1→2: compression
2→3: heat rejection
3→4: expansion
4→1: heat addition

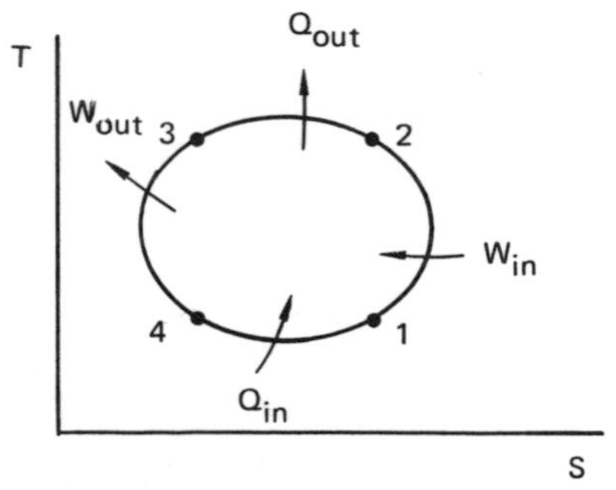

REFRIGERATION UNITS

$$1 \text{ ton} = 200 \text{ BTU/min}$$

HEAT PUMPS

A heat pump operates on a refrigeration cycle using refrigeration equipment. The only difference is the use to which a heat pump is put. The purpose of a heat pump is to warm the region in which the heat rejection coils are located.

COEFFICIENT OF PERFORMANCE

Efficiencies are not calculated for refrigeration cycles. Rather, coefficients of performance (COP) are used to compare different cycles.

$$COP_{refrigerator} = \frac{Q_{absorbed}}{W_{compression}} = \frac{Q_{absorbed}}{Q_{rejected} - Q_{absorbed}}$$

$$COP_{heat\ pump} = \frac{Q_{rejected}}{W_{compression}} = \frac{Q_{absorbed} + W_{compression}}{W_{compression}} = COP_{refrigerator} + 1$$

CARNOT CYCLE COP

$$COP_{refrigerator} = \frac{T_{low}}{T_{high} - T_{low}}$$

$$COP_{heat\ pump} = \frac{T_{high}}{T_{high} - T_{low}} = COP_{refrigerator} + 1$$

COMPRESSIBLE FLUID DYNAMICS

GENERAL FLOW EQUATION

$$\frac{v_1^2}{2g_c} + Jh_1 = \frac{v_2^2}{2g_c} + Jh_2 \qquad (\Delta z \approx 0)$$

SPEED OF SOUND

$$c = \sqrt{kg_cRT}$$

MACH NUMBER

$$M = \frac{v}{c}$$

ISENTROPIC FLOW PARAMETERS

$$\left(\frac{T_T}{T_2}\right) = \tfrac{1}{2}(k-1)M_2^2 + 1$$

$$\left(\frac{p_T}{p}\right) = \left[\tfrac{1}{2}(k-1)M^2 + 1\right]^{\frac{k}{(k-1)}}$$

$$\left(\frac{\rho_T}{\rho}\right) = \left[\tfrac{1}{2}(k-1)M^2 + 1\right]^{\frac{1}{(k-1)}}$$

$$\left(\frac{A}{A^*}\right) = \frac{1}{M}\left[\frac{\tfrac{1}{2}(k-1)M^2 + 1}{\tfrac{1}{2}(k-1) + 1}\right]^{\frac{k+1}{2(k-1)}}$$

$$\left(\frac{v}{c^*}\right) = \sqrt{\frac{\tfrac{1}{2}(k+1)M^2}{\tfrac{1}{2}(k-1)M^2 + 1}}$$

CRITICAL PRESSURE RATIO

$$R_{cp} = \left(\frac{2}{k+1}\right)^{\frac{k}{k-1}}$$

ROCKET PERFORMANCE

$$\dot{m} = \rho_e v_e A_e$$

thrust

$$F = \frac{\dot{m}\, v_e}{g_c} + A_e(P_e - P_a) = \frac{\dot{m}\, v_{eff}}{g_c}$$

characteristic velocity

$$v_{char} = \frac{g_c p_T A_t}{\dot{m}} = \frac{v_{eff}}{C_f}$$

coefficient of thrust

$$C_f = \frac{F}{p_T A_t}$$

total impulse

$$I = Ft$$

STEAM FLOW (NOZZLE FLOW)

$$v_2 = \sqrt{2g_cJ(h_1 - h_2) + v_1^2}$$

$$\eta_{nozzle} = \frac{\Delta h_{actual}}{\Delta h_{ideal}} = \left(\frac{v_{actual}}{v_{ideal}}\right)^2$$

ISENTROPIC FLOW FACTORS FOR k = 1.4

M = Mach number, P = pressure, TP = total pressure, D = density, TD = total density, T = temperature, TT = total temperature, A = area at a point, A^* = throat area for M = 1, v = velocity, and c^* = speed of sound at throat.

M	P/TP	D/TD	T/TT	A/A^*	V/C^*
.00	1.0000	1.0000	1.0000	-------	.0000
.01	.9999	1.0000	1.0000	57.8737	.0110
.02	.9997	.9998	.9999	28.9420	.0219
.03	.9994	.9996	.9998	19.3005	.0329
.04	.9989	.9992	.9997	14.4815	.0438
.05	.9983	.9988	.9995	11.5914	.0548
.06	.9975	.9982	.9993	9.6659	.0657
.07	.9966	.9976	.9990	8.2915	.0766
.08	.9955	.9968	.9987	7.2616	.0876
.09	.9944	.9960	.9984	6.4613	.0985
.10	.9930	.9950	.9980	5.8218	.1094
.11	.9916	.9940	.9976	5.2992	.1204
.12	.9900	.9928	.9971	4.8643	.1313
.13	.9883	.9916	.9966	4.4969	.1422
.14	.9864	.9903	.9961	4.1824	.1531
.15	.9844	.9888	.9955	3.9103	.1639
.16	.9823	.9873	.9949	3.6727	.1748
.17	.9800	.9857	.9943	3.4635	.1857
.18	.9776	.9840	.9936	3.2779	.1965
.19	.9751	.9822	.9928	3.1123	.2074
.20	.9725	.9803	.9921	2.9635	.2182
.21	.9697	.9783	.9913	2.8293	.2290
.22	.9668	.9762	.9904	2.7076	.2398
.23	.9638	.9740	.9895	2.5968	.2506
.24	.9607	.9718	.9886	2.4956	.2614
.25	.9575	.9694	.9877	2.4027	.2722
.26	.9541	.9670	.9867	2.3173	.2829
.27	.9506	.9645	.9856	2.2385	.2936
.28	.9470	.9619	.9846	2.1656	.3043
.29	.9433	.9592	.9835	2.0979	.3150
.30	.9395	.9564	.9823	2.0351	.3257
.31	.9355	.9535	.9811	1.9765	.3364
.32	.9315	.9506	.9799	1.9218	.3470
.33	.9274	.9476	.9787	1.8707	.3576
.34	.9231	.9445	.9774	1.8229	.3682
.35	.9188	.9413	.9761	1.7780	.3788
.36	.9143	.9380	.9747	1.7358	.3893
.37	.9098	.9347	.9734	1.6961	.3999
.38	.9052	.9313	.9719	1.6587	.4104
.39	.9004	.9278	.9705	1.6234	.4209
.40	.8956	.9243	.9690	1.5901	.4313
.41	.8907	.9207	.9675	1.5587	.4418
.42	.8857	.9170	.9659	1.5289	.4522
.43	.8807	.9132	.9643	1.5007	.4626
.44	.8755	.9094	.9627	1.4740	.4729
.45	.8703	.9055	.9611	1.4487	.4833
.46	.8650	.9016	.9594	1.4246	.4936
.47	.8596	.8976	.9577	1.4018	.5038
.48	.8541	.8935	.9560	1.3801	.5141
.49	.8486	.8894	.9542	1.3595	.5243
.50	.8430	.8852	.9524	1.3398	.5345
.51	.8374	.8809	.9506	1.3212	.5447
.52	.8317	.8766	.9487	1.3034	.5548
.53	.8259	.8723	.9468	1.2865	.5649
.54	.8201	.8679	.9449	1.2703	.5750
.55	.8142	.8634	.9430	1.2549	.5851
.56	.8082	.8589	.9410	1.2403	.5951
.57	.8022	.8544	.9390	1.2263	.6051
.58	.7962	.8498	.9370	1.2130	.6150
.59	.7901	.8451	.9349	1.2003	.6249
.60	.7840	.8405	.9328	1.1882	.6348
.61	.7778	.8357	.9307	1.1767	.6447
.62	.7716	.8310	.9286	1.1656	.6545
.63	.7654	.8262	.9265	1.1551	.6643
.64	.7591	.8213	.9243	1.1451	.6740
.65	.7528	.8164	.9221	1.1356	.6837
.66	.7465	.8115	.9199	1.1265	.6934
.67	.7401	.8066	.9176	1.1179	.7031
.68	.7338	.8016	.9153	1.1097	.7127
.69	.7274	.7966	.9131	1.1018	.7223
.70	.7209	.7916	.9107	1.0944	.7318
.71	.7145	.7865	.9084	1.0873	.7413
.72	.7080	.7814	.9061	1.0806	.7508
.73	.7016	.7763	.9037	1.0742	.7602
.74	.6951	.7712	.9013	1.0681	.7696
.75	.6886	.7660	.8989	1.0624	.7789
.76	.6821	.7609	.8964	1.0570	.7883
.77	.6756	.7557	.8940	1.0519	.7975
.78	.6691	.7505	.8915	1.0471	.8068
.79	.6625	.7452	.8890	1.0425	.8160
.80	.6560	.7400	.8865	1.0382	.8251
.81	.6495	.7347	.8840	1.0342	.8343
.82	.6430	.7295	.8815	1.0305	.8433
.83	.6365	.7242	.8789	1.0270	.8524
.84	.6300	.7189	.8763	1.0237	.8614
.85	.6235	.7136	.8737	1.0207	.8704
.86	.6170	.7083	.8711	1.0179	.8793
.87	.6106	.7030	.8685	1.0153	.8882
.88	.6041	.6977	.8659	1.0129	.8970
.89	.5977	.6924	.8632	1.0108	.9058
.90	.5913	.6870	.8606	1.0089	.9146
.91	.5849	.6817	.8579	1.0071	.9233
.92	.5785	.6764	.8552	1.0056	.9320
.93	.5721	.6711	.8525	1.0043	.9406
.94	.5658	.6658	.8498	1.0031	.9493
.95	.5595	.6604	.8471	1.0021	.9578
.96	.5532	.6551	.8444	1.0014	.9663
.97	.5469	.6498	.8416	1.0008	.9748
.98	.5407	.6445	.8389	1.0003	.9832
.99	.5345	.6392	.8361	1.0001	.9916
1.00	.5283	.6339	.8333	1.0000	1.0000

COMBUSTION

Quantities in ( ) are volumetric percents from Orsat analysis. Quantities in [ ] are gravimetric percents from ultimate analysis.

ELEMENTS AND COMPOUNDS

name	molecular weight
carbon	12.0
hydrogen	2.01
nitrogen	28.01
oxygen	32.0
sulfur	32.06

COMPOSITION OF AIR

	% by weight	% by volume
oxygen	23.15	20.9
nitrogen and inerts	76.85	79.1

HEAT OF COMBUSTION

$$HV = 14{,}093[C] + 60{,}958\left([H_2] - \frac{[O_2]}{8}\right) + 3983[S]$$

FLUE GAS ANALYSIS

$$\frac{\text{lbm actual air}}{\text{lbm fuel}} = \frac{3.04\,(N_2)[C]}{(CO_2) + (CO)}$$

$$\%\text{ excess air} = \frac{100\,((O_2) - 0.5\,(CO))}{0.264\,(N_2) - [(O_2) - 0.5\,(CO)]}$$

$$(CO_2)_{\text{theoretical}} = \frac{100\,(CO_2)_{\text{actual}}}{1 - 4.76\,(O_2)}$$

$$\frac{\text{lbm dry flue gas}}{\text{lbm solid fuel}} = \frac{\{11(CO_2) + 8(O_2) + 7\{(CO) + (N_2)\}\}\left\{[C] + \left(\frac{[S]}{1.833}\right)\right\}}{3((CO_2) + (CO))}$$

PROPERTIES OF COMMON FUELS

	Gasoline	Octane	Propane	Ethanol	No. 1 Diesel	No. 2 Diesel
Chemical formula	-	C_8H_{18}	C_3H_8	C_2H_5OH	—	—
Molecular weight	≈126	114	44	46	≈170	≈184
Carbon % by weight	—	84	82	52	—	—
Hydrogen % by weight	—	16	18	13	—	—
Oxygen % by weight	—	—	—	35	—	—
Heating value						
Higher BTU/lbm	20,260	20,590	21,646	12,800	19,240	19,110
Lower BTU/lbm	18,900	19,100	19,916	11,500	18,250	18,000
BTU/gal (lower)	116,485	111,824	81,855	76,152	133,332	138,110
Latent heat of Vaporization						
BTU/lbm	142	141	147	361	115	105
Specific gravity	.739	.702	.493	.794	.876	.920
Research octane	85–94	100	112	106		—
Motor octane	77–86	100	97	89	10–30	
Cetane number	10 to 20	—	—	−20 to 8	≈45	—
Stoichiometric Mass A/F ratio	14.7	15.1	—	9.0	—	—
Distillation Temperature (°F)	90–410	—	—	173	340–560	—
Flammability Limits (volume percent)	1.4 to 7.6	—	—	4.3 to 19	—	—

HEAT TRANSFER

THERMAL CONDUCTIVITY

$$k_T = k_o(1 + \gamma T)$$

CONDUCTION THROUGH SLABS

$$q = \frac{kA\Delta T}{L}$$

CONDUCTION THROUGH SANDWICHES

$$q = \frac{A\Delta T}{\Sigma\left(\frac{L_i}{k_i}\right)} \qquad \text{(no films)}$$

$$q = \frac{A\Delta T}{\Sigma\left(\frac{L_i}{k_i}\right) + \Sigma\left(\frac{1}{h_j}\right)} \qquad \text{(films)}$$

COEFFICIENT OF HEAT TRANSFER (CONDUCTIVITY)

$$U = \frac{q}{A\Delta T} \qquad \text{(general rule)}$$

$$U = \frac{1}{\Sigma\left(\frac{L_i}{k_i}\right) + \Sigma\left(\frac{1}{h_j}\right)} \qquad \text{(sandwiches with films)}$$

LOGARITHMIC MEAN AREA

$$A_m = \frac{A_o - A_i}{ln\left(\frac{A_o}{A_i}\right)}$$

HOLLOW CYLINDER

$$q = \frac{2\pi kL\Delta T}{ln\left(\frac{r_o}{r_i}\right)}$$

$$A_m = \frac{2\pi l\,(r_o - r_i)}{ln\left(\frac{r_o}{r_i}\right)}$$

INSULATED PIPE

$$q = \frac{2\pi L\Delta T}{\frac{1}{r_a h_a} + \frac{ln\left(\frac{r_b}{r_a}\right)}{k_{pipe}} + \frac{1}{r_b h_b} + \frac{ln\left(\frac{r_c}{r_b}\right)}{k_{in}} + \frac{1}{r_c h_c}}$$

note: h_b may not be present

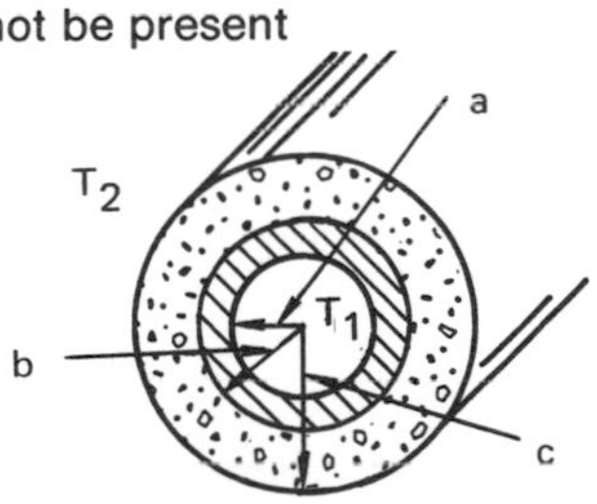

CRITICAL THICKNESS OF INSULATION

$$r_{critical} = \frac{k_{insulation}}{h}$$

NEWTON'S LAW OF COOLING

$$T_t = T_\infty + (T_o - T_\infty)e^{-rt}$$

$$t = \left(\frac{1}{r}\right) ln\left(\frac{T_t - T_\infty}{T_o - T_\infty}\right)$$

The rate constant, *r*, must be determined from heat loss data.

BIOT NUMBER

$$N_{Bi} = \frac{hL}{k}$$

THERMAL DIFFUSIVITY

$$a = \frac{k}{\rho c_p}$$

FOURIER NUMBER

$$N_{Fo} = \frac{kt}{\rho c_p L^2} = \frac{at}{L^2}$$

TRANSIENT HEAT FLOW

(only for $N_{Bi} < 10$)

$$T_t = T_\infty + \Delta T e^{-N_{Bi} N_{Fo}}$$

$$q_t = hA_s \Delta T e^{-N_{Bi} N_{Fo}}$$

$$\Delta T = T_{initial} - T_{environment}$$

REYNOLDS' NUMBER

$$N_{Re} = \frac{Dv\rho}{\mu} = \frac{Dv}{\nu}$$

PRANDTL NUMBER

$$N_{Pr} = \frac{C_p\mu}{k}$$

GRASHOFF NUMBER

$$N_{Gr} = \frac{L^3\rho^2\beta\Delta Tg}{\mu^2}$$

$$\Delta T = T_{surface} - T_{environment}$$

Evaluate film properties at $\frac{1}{2}(T_{surface} + T_{environment})$

NUSSELT EQUATION FOR FORCED CONVECTION IN PIPES

$$\frac{hD}{k} = .0225\,(N_{Re})^{.8}\,(N_{Pr})^n$$

$n = .3$ for heat flow out of pipe
$= .4$ for heat flow into pipe

NATURAL CONVECTION

$$\frac{hL}{k} = C\,(N_{Gr}N_{Pr})^n \text{ for } N_{Pr} > .6$$

C and *n* depend on the configuration.

Film properties are evaluated at $\frac{1}{2}(T_{surface} + T_{environment})$

LOGARITHMIC MEAN TEMPERATURE DIFFERENCE

$$\Delta T_m = \frac{\Delta T_A - \Delta T_B}{ln\left(\frac{\Delta T_A}{\Delta T_B}\right)}$$

ΔT_A *and* ΔT_B are the temperature differences between fluids at ends A and B. (Assume counterflow operation for multiple pass and crossflow exchangers.)

HEAT EXCHANGERS

$$q = UA\Delta T_m \qquad \text{(single pass)}$$

$$q = F_c UA\Delta T_m \qquad \text{(multiple pass)}$$

$$\frac{1}{U} \approx \frac{1}{h_o} + \frac{1}{h_i} + R_{fo} + R_{fi} \quad \text{(with fouling factors)}$$

F_c is 1.0 if either fluid is constant in temperature (e.g., changing phase).

STEFAN-BOLTZMAN CONSTANT

$$\sigma = .1713 \text{ EE} - 8 \frac{\text{BTU}}{\text{hr-ft}^2\text{-}^\circ R^4}$$

RADIATION FROM A GREY BODY

$$E_g = \epsilon_g \sigma T^4$$

ϵ is the emissivity.
T is in °R.

NET RADIATION TRANSFER BETWEEN TWO GREY BODIES

$$E = \sigma F_e F_a [(T_1)^4 - (T_2)^4]$$
$$= \sigma F_{12} [(T_1)^4 - (T_2)^4]$$

$F_{12} = \epsilon_{inner}$ for completely enclosed bodies and infinite parallel planes.

HEATING, VENTILATING, AND AIR CONDITIONING

(Q is in ft³/hr. ω is in lbm/lbm. q is in BTU/hr.)

HEATING LOAD

HEAT LOSSES THROUGH WALLS

$$q = UA\,(T_i - T_o)$$

INFILTRATION HEAT LOSSES

$$q_a = 0.24\,Q\rho\,(T_i - T_o)$$
$$\approx 0.018\,Q\,(T_i - T_o)$$

HEAT TO WARM MOISTURE

$$q_w = m_w c_p \Delta T = \rho Q \omega c_p\,(T_i - T_o)$$
$$\approx (.075)\,(Q)\,(\omega)\,(.45)\,(T_i - T_o)$$
$$\approx (.0338)\omega Q\,(T_i - T_o)$$

TOTAL HEAT TO WARM INCOMING AIR

$$q_t = q_a + q_w$$
$$= m_a\,(h_i - h_o)$$

LATENT HEAT TO ADD MOISTURE

$$q = h_{fg} Q \rho (\omega_i - \omega_o) \approx 79.5 Q (\omega_i - \omega_o)$$

HEAT CONTRIBUTED BY LIGHTS AND MACHINES

$$q_{\text{motors}} = \frac{(2545)(hp)}{\eta}$$

$$q_{\text{lights}} = (3413)\,(kw)$$

DEGREE DAYS

$$DD = N(65 - T_{\text{ave}})$$

N is the number of days in the heating season.

ANNUAL FUEL CONSUMPTION

$$\text{fuel consumption} = \frac{24q(DD)}{(T_i - T_o)(HV)\eta_c}$$

HV is the fuel heating value.

VENTILATING

15 cfm of fresh air per person is commonly used as a design standard. In areas with excessive activity and smoking, this should be increased to 25-40 cfm.

AIR CONDITIONING PROCESSES

SENSIBLE COOLING

$$q_{\text{removed}} = m_a(h_1 - h_2) \approx m_a(.24 + .45\omega)(T_1 - T_2)$$

$$\text{BF} = \text{bypass factor} = \frac{T_{2,db} - T_{\text{coil}}}{T_{1,db} - T_{\text{coil}}}$$

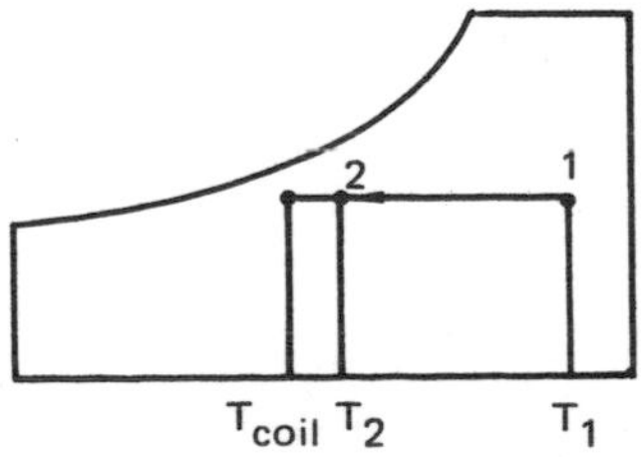

SENSIBLE HEATING

$$q_{\text{added}} = m_a(h_2 - h_1) \approx m_a(.24 + .45\omega)\,(T_2 - T_1)$$

$$\text{BF} = \frac{T_{\text{coil}} - T_2}{T_{\text{coil}} - T_1}$$

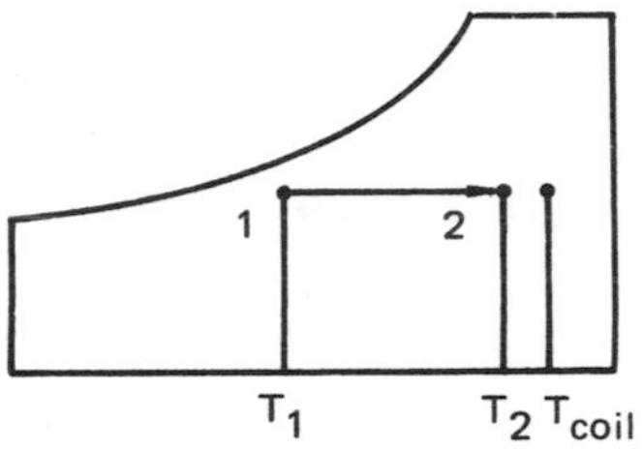

ADIABATIC MIXING OF TWO AIR STREAMS

h and ω are linear scales on the psychrometric chart. Therefore, it is possible to use straight line proportions in the same ratio as m_{a1}/m_{a2}. To do so, draw a straight line between the two points (1 and 2) representing the input streams. The final mixture (3) will be on that line. Use the *lever rule* (on the basis of air masses) to locate the mixture point. Since the water vapor adds little to the mixture volume, the air volumes can be approximated by the total mixture volumes.

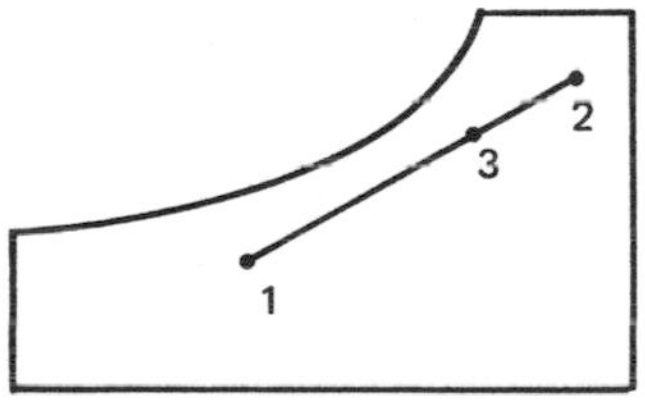

COOLING AND COIL DEHUMIDIFICATION

For convenience, it is assumed that a straight line can be drawn between point 1 and the ADP, and that the final condition of the air will be along that straight line. Point 2 can be located if the coil bypass factor is known.

$$BF = \frac{T_{db,2} - ADP}{T_{db,1} - ADP} = \frac{\text{line segment } 4 - 2}{\text{line segment } 4 - 1}$$

$$q_t = m_a(h_1 - h_2)$$
$$q_l = m_a(\omega_1 - \omega_2)h_{fg}$$
$$q_s = q_t - q_l$$

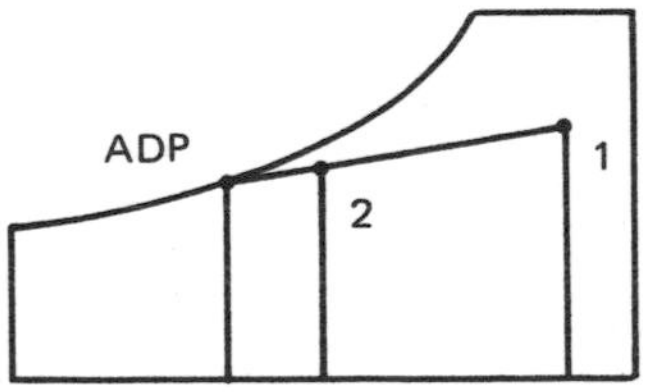

COOLING BY ADIABATIC SATURATION (EVAPORATIVE COOLING)

$$\eta_{\text{sat}} = \frac{T_{a,\text{out}} - T_w}{T_{a,\text{in}} - T_w}$$

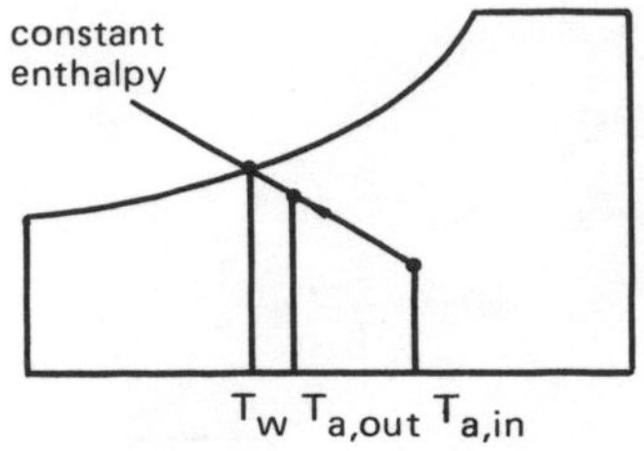

COOLING LOAD

INSTANTANEOUS HEAT GAIN THROUGH WALLS

$$q = UA\Delta T_{te}$$

ΔT_{te} is the total equivalent temperature difference.

HEAT GAIN THROUGH WINDOWS

$$q = A[C_s F_{shg} + U(T_o - T_i)]$$

C_s is the shading coefficient.

F_{shg} is the solar heat gain factor.

HEAT GAIN FROM LIGHTS AND EQUIPMENT

$$q = 3.41 \text{ (wattage)}$$

$$q = \frac{2545 \text{ (horsepower)}}{\eta}$$

ENERGY EFFICIENCY RATIO

$$EER = \frac{\text{cooling (BTU/hr)}}{\text{input power (watts)}}$$

PSYCHROMETRIC CHART

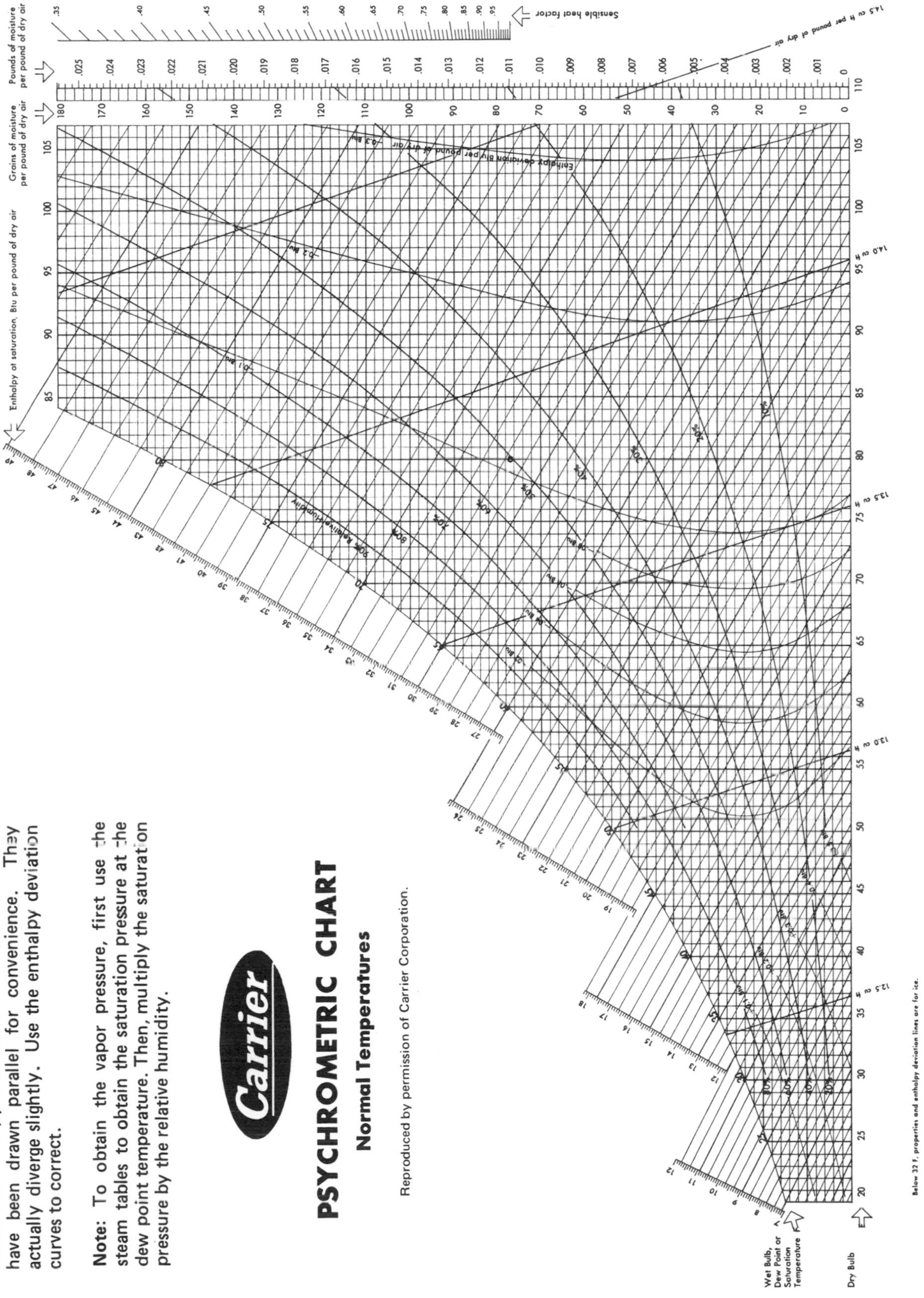

Note: The enthalpy and wet bulb temperature scales have been drawn parallel for convenience. They actually diverge slightly. Use the enthalpy deviation curves to correct.

Note: To obtain the vapor pressure, first use the steam tables to obtain the saturation pressure at the dew point temperature. Then, multiply the saturation pressure by the relative humidity.

Carrier

PSYCHROMETRIC CHART

Normal Temperatures

STATICS

FORCES

A force is a vector quantity. As such, it is completely defined by giving its magnitude, line of application, sense, and point of application. Units of force are pounds in the English system and newtons in the SI system.

RESULTANTS

The resultant of n forces with components $F_{x,i}$ and $F_{y,i}$ has a magnitude of

$$R = \sqrt{(\Sigma F_{x,i})^2 + (\Sigma F_{y,i})^2}$$

The direction is

$$\phi = \arctan\left[\Sigma F_{y,i}/\Sigma F_{x,i}\right]$$

COUPLES

A couple is a moment created by two equal forces acting parallel but with opposite directions. The effect of a couple is to cause rotation. If d is the separation distance then

$$M = Fd$$

FORCES IN VECTOR FORM

A force R may be separated into its components by using the direction cosines.

$$F_x = R\cos\theta_x$$

$$F_y = R\cos\theta_y$$

$$F_z = R\cos\theta_z$$

$$\cos\theta_x = x/\sqrt{x^2 + y^2 + z^2}$$

$$\cos\theta_y = y/\sqrt{x^2 + y^2 + z^2}$$

$$\cos\theta_z = z/\sqrt{x^2 + y^2 + z^2}$$

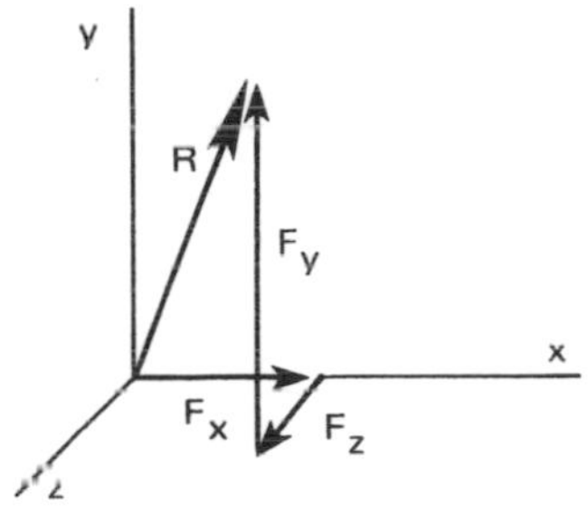

The resultant may be written in vector form in terms of its components and the unit vectors. (The addition is vector addition, not algebraic addition.)

$$R = iF_x + jF_y + kF_z$$

COMPONENTS OF INCLINED MEMBERS

A non-trigonometric method can be used to resolve forces in inclined members into their components. This resolution can be accomplished by multiplying the force by the ratio of sides, as determined from the geometry of the inclined member.

$$F_y = (y/h)F$$

$$F_x = (x/h)F$$

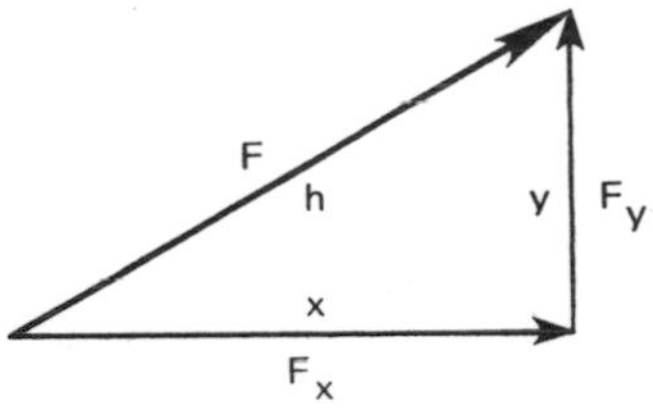

MOMENTS

The moment produced by a force acting with a moment arm of length d is

$$M = (F)(d)$$

CONDITIONS FOR EQUILIBRIUM

In general, the conditions for equilibrium are:

$$\Sigma F_x = 0,\ \Sigma F_y = 0,\ \Sigma F_z = 0,\ \Sigma M_x = 0,\ \Sigma M_y = 0,\ \Sigma M_z = 0$$

For general coplanar systems, the conditions are:

$$\Sigma F_x = 0,\ \Sigma F_y = 0,\ \Sigma M. = 0$$

For parallel systems, the conditions are:

$$\Sigma F_x = 0,\ \Sigma M_z = 0$$

For concurrent systems, the conditions are:

$$\Sigma F_x = 0,\ \Sigma F_y = 0$$

SIGN CONVENTIONS FOR TRUSS FORCES

For forces in truss members, tension is positive and compression is negative. When freebodies are drawn for truss *joints*, forces leaving the joints place the truss member in tension; forces going into the joints place the member in compression.

DETERMINATE TRUSSES

A truss will be determinate if the number of truss members is equal to

no. truss members = 2(no. of joints) - 3

If the left-hand side is less than the right-hand side, the truss is not rigid. If the left-hand side is greater than the right-hand side, the truss is indeterminate to the degree of the difference.

CATENARY CABLES

In the figure and equations below, c is the distance from the lowest point on the cable to a reference plane below. The distance c is a constant which must be determined. It does not correspond to any physical dimension.

$$y = c\,(\cosh(x/c))$$

$$s = c\,(\sinh(x/c))$$

$$y = \sqrt{s^2 + c^2}$$

$$S = c\,(\cosh(a/c) - 1)$$

$$\tan\theta = s/c$$

$H = wc$	(horizontal component of tension)
$F = ws$	(applied vertical load due to cable weight)
$T = wy$	(tangential tension)

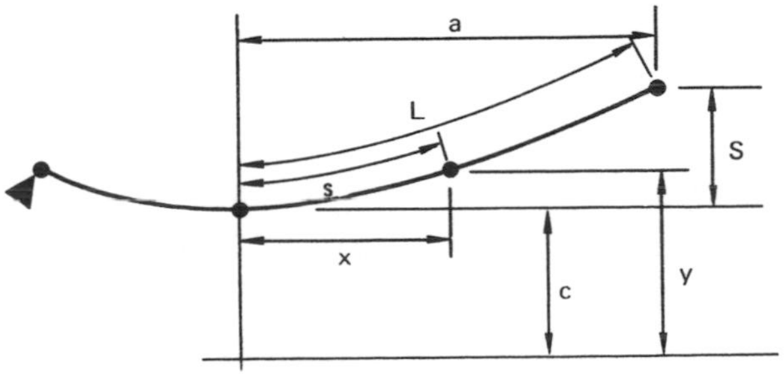

FRICTION

The frictional force depends on the coefficient of friction (μ) and the normal force (N).

$$F_f = \mu N$$

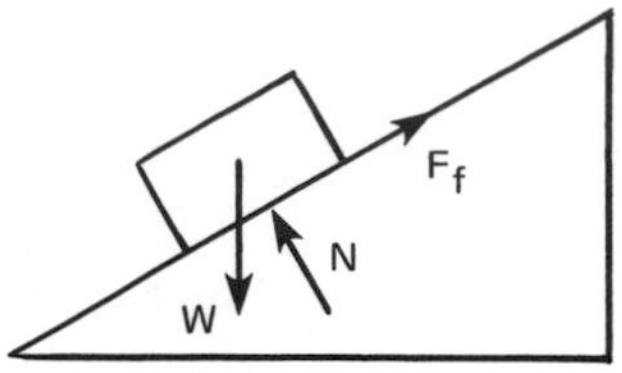

CENTROIDS

The centroid of an object is that point at which the object would balance if suspended. In general, the *x* and *y* coordinates of the centroidal location can be found from the following relationships:

$$x_c = (1/A)\int x \, dA$$

$$y_c = (1/A)\int y \, dA$$

The centroidal locations of common shapes are given below:

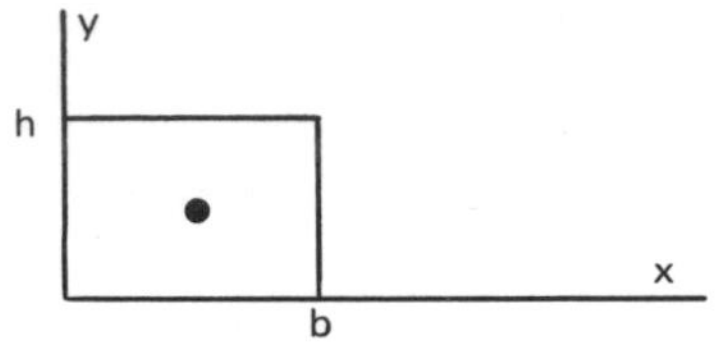

$x_c = \frac{1}{2}b$

$y_c = \frac{1}{2}h$

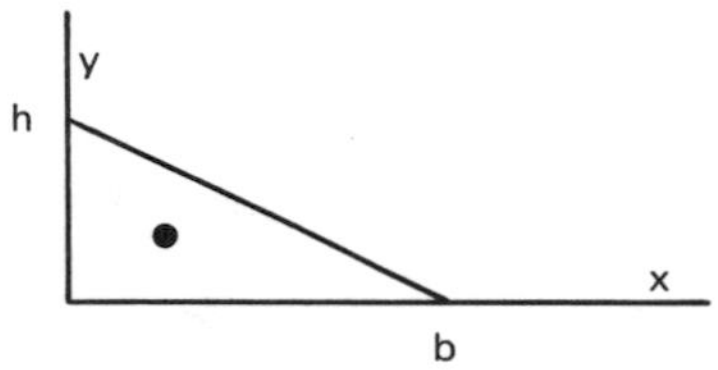

$x_c = (\frac{1}{3})b$

$y_c = (\frac{1}{3})h$

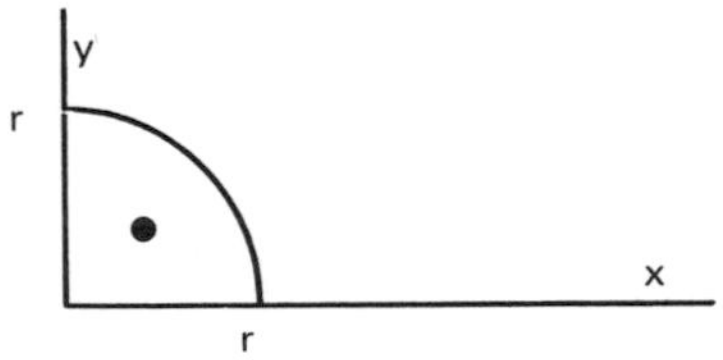

$x_c = 4r/3\pi$

$y_c = 4r/3\pi$

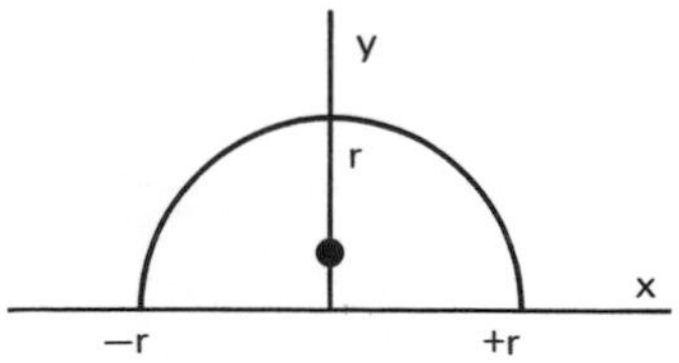

$x_c = 0$

$y_c = 4r/3\pi$

CENTROIDS OF COMPOSITE (BUILT-UP) SHAPES

$$x_c = \Sigma A_i x_{c,i} / \Sigma A_i$$

$$y_c = \Sigma A_i y_{c,i} / \Sigma A_i$$

MOMENTS OF INERTIA

The moment of inertia (second moment of area) has units of (length)^4 and can be considered a measure of resistance to bending. I_x and I_y represent the resistance to bending about the *x* and *y* axes respectively. I_x and I_y are not components.

$$I_x = \int y^2 dA$$

$$I_y = \int x^2 dA$$

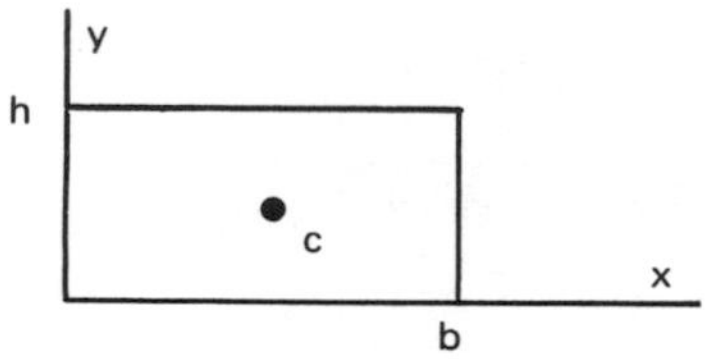

$I_{x,c} = (1/12)bh^3$

$I_x = (1/3)bh^3$

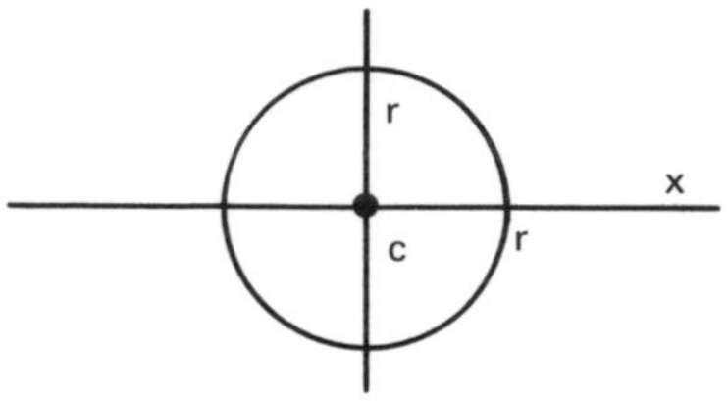

$I_{x,c} = \frac{1}{4}\pi r^4$

RADIUS OF GYRATION

The radius of gyration is the distance from a reference axis at which all of the area can be considered to be concentrated to produce the actual moment of inertia.

$k = \sqrt{I/A}$ or $k = \sqrt{J/A}$ for polar moments of inertia

POLAR MOMENTS OF INERTIA

In general, the polar moment of inertia is

$$J = \int r^2 dA = I_x + I_y$$

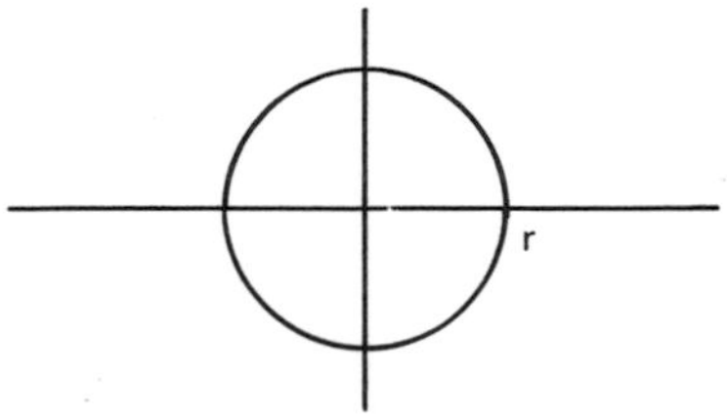

$J = \frac{1}{2}\pi r^4 = (1/32)\pi d^4$

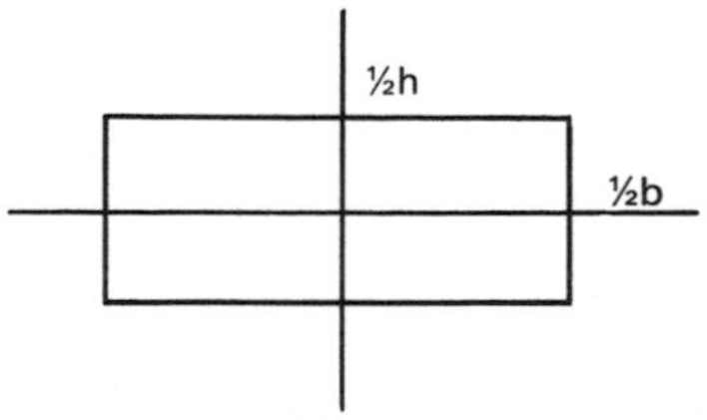

$J = (1/12)bh(b^2 + h^2)$

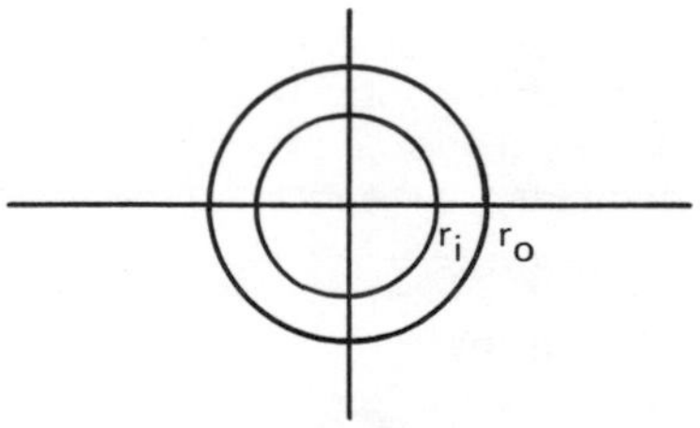

$J = \frac{1}{2}\pi(r_o^4 - r_i^4)$

PARALLEL AXIS THEOREM

If the moment of inertia is known about the centroidal axis, the moment of inertia about a second parallel axis located a distance of *d* away from the centroidal axis is:

$$I_{new} = I_{centroidal} + Ad^2$$

MATERIALS SCIENCE

TENSILE TEST

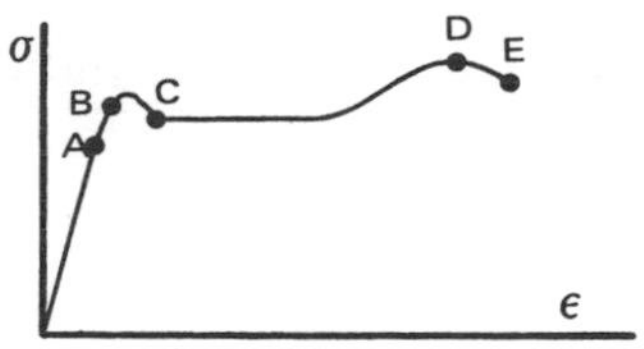

engineering stress: $\sigma = P/A_o$

true stress: $\Sigma = P/A_{instantaneous}$

engineering strain: $\epsilon = \Delta L/L_o$

true strain: $E = \ln(A_o/A_{instantaneous})$

point A: proportionality limit - the highest stress for which Hooke's law ($\sigma = E\epsilon$) is valid.

point B: elastic limit - the highest stress for which no permanent deformation occurs.

point C: yield point - the stress at which a sharp drop in load-carrying ability occurs.

point D: ultimate strength - the highest stress which the material can achieve.

point E: fracture strength - the stress at fracture.

The *modulus of elasticity* (E) is the slope of the line for stresses up to the proportionality limit.

The *shear modulus* (G) can be calculated from the modulus of elasticity and Poisson's ratio.

$$G = E/2(1 + \mu)$$

The *toughness* is the work per unit volume required to cause fracture. It is calculated as the area under the σ-ϵ curve up to the point of fracture. Units of toughness are in-lbf/in³.

The *ductility* is a measure of the amount of plastic strain at the breaking point. It is calculated as the per cent reduction in area at fracture.

$$\text{reduction in area} = (A_o - A_{frac.})/A_o = (L_{frac.} - L_o)/L_o$$

The *per cent elongation at fracture* is calculated from the original length and the fracture length after the sample has 'snapped back.'

ENDURANCE TEST

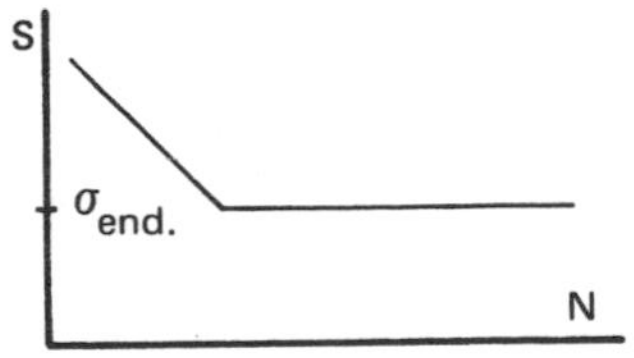

Endurance tests (fatigue tests) apply a cyclical loading of constant maximum amplitude. The plot (usually semi-log or log-log) of the maximum stress and the number of cycles to failure is known as an S-N plot.

The *endurance stress (endurance limit or fatigue limit)* is the maximum stress which can be repeated indefinitely without causing failure.

The *fatigue life* is the number of cycles required to cause failure for a given stress level.

IMPACT TEST

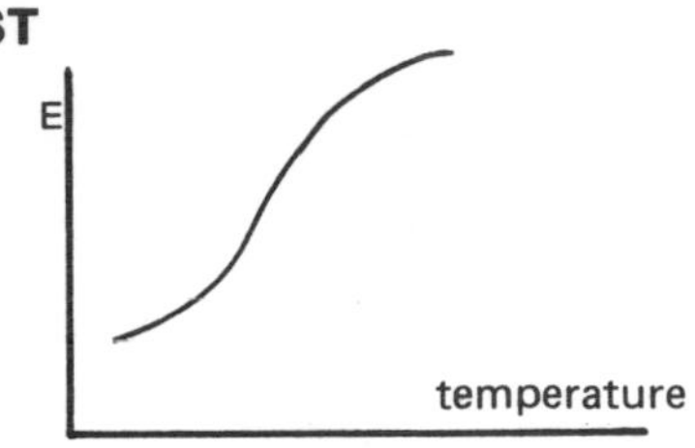

Impact tests determine the amount of energy required to cause failure in standardized test samples. The tests are repeated over a range of temperatures to determine the *transition temperature.* (The transition temperature is approximately 32°F for low-carbon steel.)

CREEP TEST

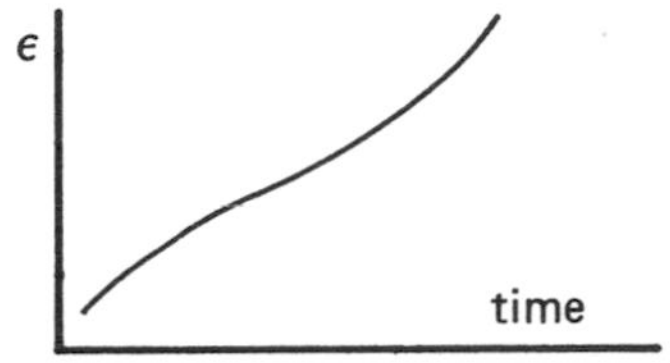

A constant stress less than the yield strength is applied and the elongation versus time is measured.

The *creep rate* ($d\epsilon/dt$) is very temperature dependent.

The *creep strength* is the stress which results in a given creep rate.

The *rupture strength* is the stress which results in failure after some given amount of time.

HARDNESS VERSUS ULTIMATE STRENGTH

$$S_{ut} \approx (500)(BHN)$$

STEEL ALLOYING INGREDIENTS

(XX is the carbon content, 0.XX%)

alloy number	major alloying elements
10XX	plain carbon steel
11XX	resulfurized plain carbon
13XX	manganese
23XX, 25XX	nickel
31XX, 33XX	nickel, chromium
40XX	molybdenum
41XX	chromium, molybdenum
43XX	nickel, chromium, molybdenum
46XX, 48XX	nickel, molybdenum
51XX	chromium
61XX	chromium, vanadium
81XX, 86XX, 87XX	nickel, chromium, molybdenum
92XX	silicon

ALUMINUM ALLOYING INGREDIENTS

alloy number	major alloying ingredient
1XXX	commercially pure aluminum (99+%)
2XXX	copper
3XXX	manganese
4XXX	silicon
5XXX	magnesium
6XXX	magnesium and silicon
7XXX	zinc
8XXX	other

ALUMINUM TEMPERS

temper	description
T2	annealed (castings only)
T3	solution heat-treated, followed by cold working
T4	solution heat-treated, followed by natural aging
T5	artificial aging
T6	solution heat-treated, followed by artificial aging
T7	solution heat-treated, followed by stabilizing by overaging heat treating
T8	solution heat-treated, followed by cold working and subsequent artificial aging

GALVANIC SERIES

(Anodic to Cathodic)

Magnesium alloys
Alclad 3S
Aluminum alloys
Low-carbon steel
Cast iron
Stainless—No. 410
Stainless—No. 430
Stainless—No. 404
Stainless—No. 316
Hastelloy A
Lead-tin alloys
Brass
Copper
Bronze
90/10 Copper-nickel
70/30 Copper-nickel
Inconel
Silver
Stainless steels (passive)
Monel
Hastelloy C
Titanium

PREVENTING CORROSION

In many designs, corrosion can be reduced or eliminated entirely by avoiding conditions conducive to corrosion. Several design principles for avoiding corrosion are available.

- Metals should be chosen on the basis of their potential for corrosion. This requires attention to chemical properties and environment. When two different metals must be in contact, they should be as close together in the galvanic series as possible.
- Protective coatings (e.g., sodium silicate, sodium benzoate, and various organic amines) can be used.
- Dampness should be eliminated. If there is no electrolyte, there can be no corrosion.
- Since each metal exhibits maximum corrosion at a particular pH, acidity or alkalinity of the environment can be controlled.
- An electrical current can be applied to the corrosion circuit to counter the corrosion reaction. This is known as *cathodic protection*.
- *Sacrificial anodes* made from active metals (e.g., magnesium or calcium) can be placed in areas where corrosion would otherwise occur. These anodes intercept the corrosive current (i.e., they give off electrons) and are sacrificed in the process of saving the structure.

MECHANICS OF MATERIALS

STRESS

Normal stress: $\sigma = \frac{F}{A}$

Shear stress: $\tau = \frac{F}{A}$

STRAIN

$$\epsilon = \frac{\Delta L}{L_o}$$

HOOKE'S LAW

For normal stress: $\sigma = E\epsilon$

For shear stress: $\tau = G\epsilon_z$

ELONGATION UNDER NORMAL STRESS

$$\delta = L_o\epsilon = \frac{FL_o}{AE}$$

VALUES OF E AND G

Steel: E = 3 EE7 psi (hard), 2.9 EE7 psi (soft); G = 1.15 EE7 psi

Aluminum: E = 1 EE7 psi; G = 3.85 EE6 psi

POISSON'S RATIO

μ is the ratio of traverse strain to longitudinal strain.

$$\mu = \frac{\Delta D L_o}{D_o \Delta L}$$

Typical values of μ are .3 for steel and .33 for aluminum.

THERMAL STRESS AND STRAIN

The elongation of an object when heated is

$$\Delta L = \alpha L_o \Delta T$$

Values of α are 6 EE-6 in/in-°F for steel and 1.3 EE-5 in/in-°F for aluminum.

The thermal stress and strain are:

$$\epsilon_{th} = \alpha \Delta T$$

$$\sigma_{th} = \epsilon_{th} E$$

NORMAL STRESS IN A BEAM

Normal stress in a beam is also called 'flexure stress' and 'bending stress'.

$$\sigma = \frac{My}{I}$$

$$\sigma_{max} = \frac{Mc}{I} = \frac{M}{Z}$$

$I = \frac{bh^3}{12}$ for a rectangle

Z = section modulus $= \frac{I}{c}$

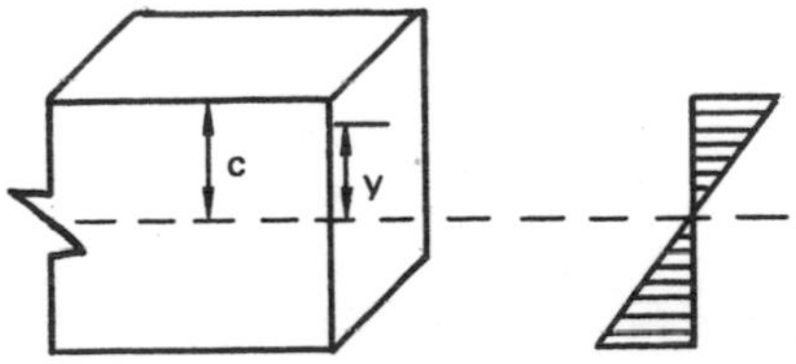

SHEAR STRESS IN A BEAM

$$\tau = \frac{VQ}{Ib}$$

$\tau_{max} = \frac{3V}{2A}$ (rectangle)

$= \frac{4V}{3A}$ (circular)

V = shear (pounds)

Q = statical moment, also known as 'first moment of the area'

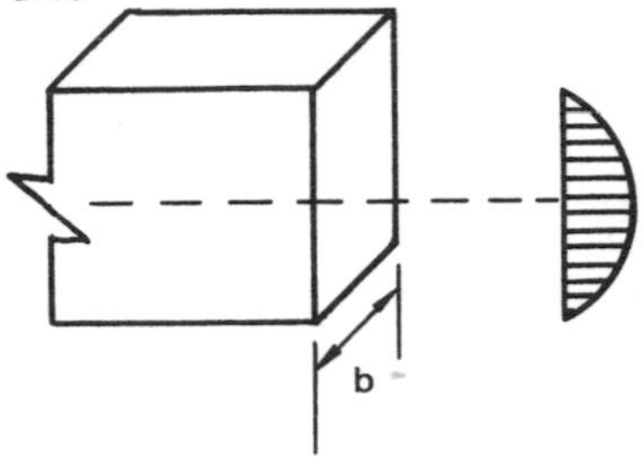

TORSIONAL STRESS AND STRAIN IN A SHAFT

$$\tau = \frac{Tc}{J}$$

J = polar (area) moment of inertia

$= \frac{1}{2}\pi r^4$ (round shafts)

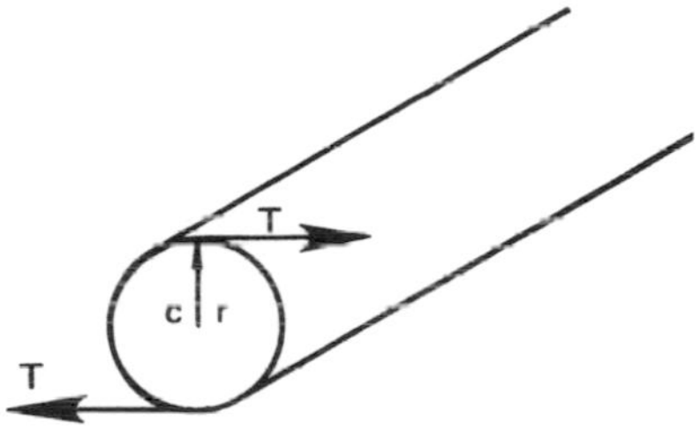

Angle of twist $= \frac{TL}{GJ}$ (in radians)

The maximum torque that the shaft can carry is

$$T_{max} = \frac{\tau_{max} J}{r}$$

The horsepower transmitted by the shaft is

$hp = \frac{2\pi T(\text{rpm})}{33{,}000}$ (T in ft-lbs)

ECCENTRIC NORMAL STRESS

$$\sigma_{max} = \left(\frac{F}{A}\right) \pm \frac{Mc}{I}$$

$$M = Fe$$

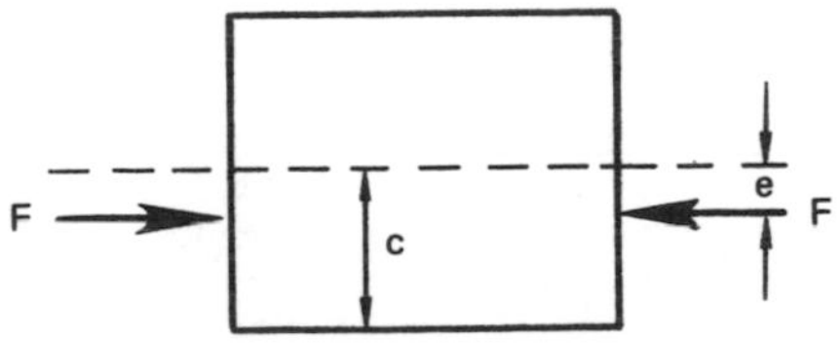

COMBINED STRESS

The normal and shear stresses on a plane whose normal is inclined an angle θ from the horizontal are

$$\sigma_\theta = \tfrac{1}{2}(\sigma_x + \sigma_y) + \tfrac{1}{2}(\sigma_x - \sigma_y)\cos 2\theta + \tau \sin 2\theta$$

$$\tau_\theta = -\tfrac{1}{2}(\sigma_x - \sigma_y)\sin 2\theta + \tau \cos 2\theta$$

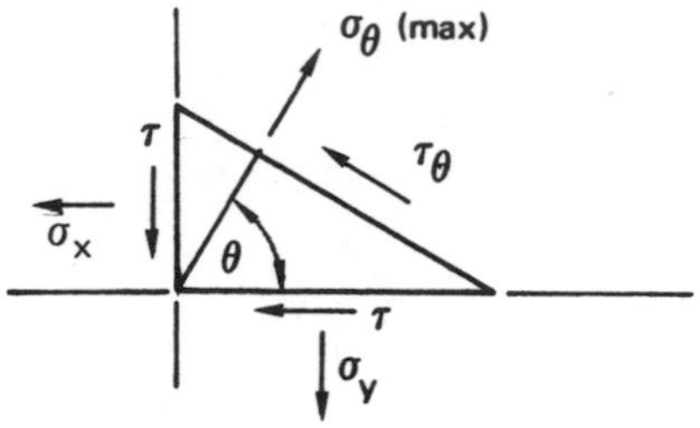

The maximum and minimum values of σ_θ and τ_θ (as θ is varied) are the principal stresses. These are

$$\sigma(\text{max, min}) = \tfrac{1}{2}(\sigma_x + \sigma_y) \pm \tau(\text{max})$$

$$\tau(\text{max, min}) = \pm \tfrac{1}{2}\sqrt{(\sigma_x - \sigma_y)^2 + (2\tau)^2}$$

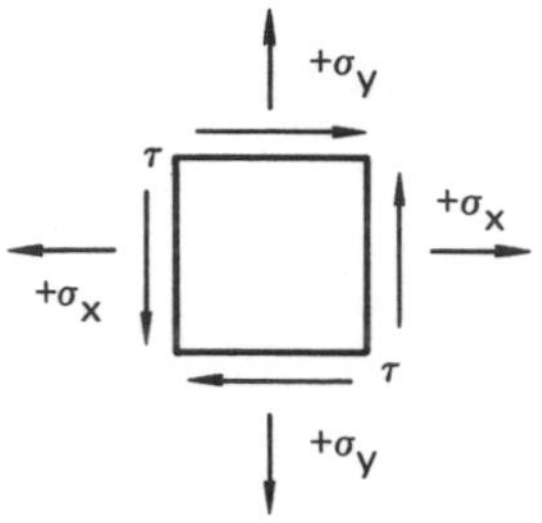

Proper sign convention must be adhered to. Normal tensile stresses are positive; normal compressive stresses are negative. Shear stresses are positive as shown.

ALLOWABLE STRESS

The allowable stress may be calculated from either the yield stress (typical for ductile materials like steel) or the ultimate strength (typical for brittle materials like cast iron).

$$\sigma_{\text{allowable}} = S_{\text{yield}}/\text{factor of safety} \qquad \text{(ductile)}$$

$$= S_{\text{ultimate}}/\text{factor of safety} \qquad \text{(brittle)}$$

MOMENT DIAGRAMS

- Clockwise moments are positive. (Use the left-hand rule.)
- Concentrated loads produce straight inclined lines.
- Uniform loads produce parabolic lines.
- Maximum moment occurs where shear (V) is zero.
- Moment is zero at a free end or hinge.
- Moment at any point is the area under the shear diagram up to that point. That is, $M = \int V\,dx$.

SHEAR DIAGRAMS

- Loads and reactions acting up are positive.
- Shear at any point is the sum of forces up to that point.
- Concentrated loads produce horizontal straight lines.
- Uniform loads product straight inclined lines.
- Shear at any point is the slope of the moment diagram at that point. That is, $V = dM/dx$.

SIMPLE BEAM DEFLECTIONS

type of beam	loading	M_{max}	deflection
Cantilever	at tip	FL	$FL^3/3EI$
Cantilever	uniform	$\tfrac{1}{2}wL^2$	$wL^4/8EI$
Simple	at center	$\tfrac{1}{4}FL$	$FL^3/48EI$
Simple	uniform	$wL^2/8$	$5wL^4/384EI$

SLENDER COLUMNS IN COMPRESSION

Euler's formula may be used if the actual stress is kept less than the yield strength. Slender columns fail by buckling in a smooth curve. A slender column has a slenderness ratio of approximately 80 or higher.

$$\text{slenderness ratio} = \frac{L}{k}$$

$$F_{\text{buckling}} = \frac{\pi^2 EI}{(CL)^2\,(\text{factor of safety})}$$

end conditions	C
both ends pinned	1
both ends built in	.5
one end pinned, one end built in	.707
one end built in, one end free	2

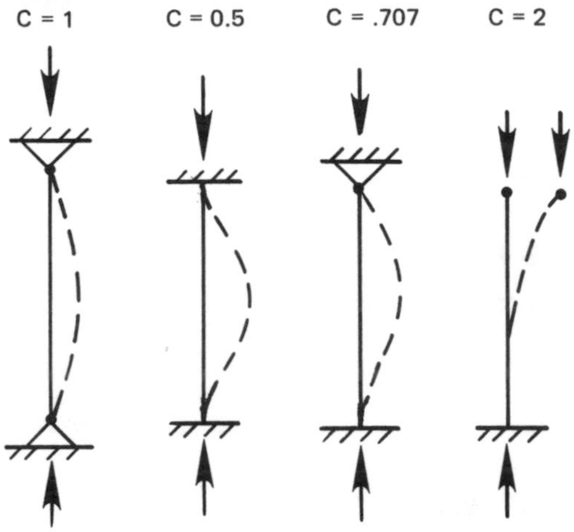

SPRINGS

$$F = kx$$

$$W = \tfrac{1}{2}kx^2$$

$$k = k_1 + k_2 + k_3 + \ldots \qquad \text{(parallel springs)}$$

$$\frac{1}{k} = \frac{1}{k_1} + \frac{1}{k_2} + \frac{1}{k_3} + \cdots \qquad \text{(series springs)}$$

THIN-WALLED CYLINDERS

A tank has thin walls if the wall thickness is less than (1/10) of the tank diameter.

$$\sigma_{\text{hoop}} = \frac{pr}{t}$$

$$\sigma_{\text{long}} = \frac{pr}{2t}$$

The hoop and long stresses are the principal stresses. They do not combine.

For spherical tanks or spherical ends on a cylindrical tank, use the long stress formula.

SHAFTS

$$\tau = \frac{Tr}{J}$$

$$J = \frac{\pi r^4}{2} = \frac{\pi D^4}{32} \qquad \text{(round shafts)}$$

$$J = \frac{\pi}{2}\left[r_o^4 - r_i^4\right] \qquad \text{(hollow shafts)}$$

The angle of twist in radians is

$$\phi = \frac{TL}{GJ}$$

The shear modulus, if unknown, can be calculated from the modulus of elasticity and Poisson's ratio.

$$G = \frac{E}{2(1 + \mu)}$$

Horsepower, torque, and rpm are related.

$$T_{\text{in-lbf}} = \frac{(63{,}025)\ (\text{horsepower})}{\text{rpm}}$$

THICK-WALLED CYLINDERS

The circumferential (tangential) and radial stresses are maximum and minimum values at either the inner or outer surfaces.

stress	external pressure, p	internal pressure, p
σ_{co}	$\frac{-(r_o^2+r_i^2)p}{r_o^2-r_i^2}$	$\frac{2r_i^2 p}{r_o^2-r_i^2}$
σ_{ro}	$-p$	0
σ_{ci}	$\frac{-2r_o^2 p}{r_o^2-r_i^2}$	$\frac{(r_o^2+r_i^2)p}{r_o^2-r_i^2}$
σ_{ri}	0	$-p$
τ_{max}	$(\frac{1}{2})\,\sigma_{ci}$	$(\frac{1}{2})\,(\sigma_{ci}+p)$

The *diametral strain*, $\Delta D/D$, and the *circumferential strain*, $\Delta C/C$, are equal in a circular cylinder under pressure loading.

$$\frac{\Delta D}{D} = \frac{\Delta C}{C} = \frac{\Delta r}{r} = \frac{\sigma_c - \mu(\sigma_r + \sigma_L)}{E}$$

If two cylinders are pressed together with an initial interference, I, the pressure, p, acting between them expands the outer cylinder and compresses the inner one. *Interference* usually means diametral interference.

$$I = |\Delta D_i|_{\text{outer cylinder}} + |\Delta D_o|_{\text{inner cylinder}}$$

When pieces are pressed together, the assembly force can be calculated as a sliding frictional force based on the normal force.

$$F_{\text{max, assembly}} = fN = 2\pi r_{\text{shaft}} Lpf$$

The coefficient of friction for press fits is highly variable, having been reported in the range of .03 to .33.

If the hub is acted upon by a torque or a torque-causing force, the maximum resisting torque is

$$T_{max} = 2\pi r^2_{\text{shaft}} Lpf$$

MACHINE DESIGN FAILURE THEORIES

MAXIMUM SHEAR STRESS THEORY (DUCTILE MATERIALS)

Failure occurs when

$$\tau_{max} > \frac{S_{yt}}{2}$$

The factor of safety is

$$FS = \frac{S_{yt}}{2\tau_{max}}$$

The yield point in shear is predicted as 0.5 S_{yt}.

DISTORTION ENERGY AND VON MISES THEORY (DUCTILE MATERIALS)

Failure is predicted as the Von Mises stress.

$$\sigma' > S_{yt}$$

$$\sigma' = \sqrt{\sigma_1^2 + \sigma_2^2 - \sigma_1\sigma_2}$$

σ_1 and σ_2 are the principal stresses

$$FS = \frac{S_{yt}}{\sigma'}$$

The yield point in shear is predicted as 0.577 S_{yt}.

ENDURANCE STRENGTH

For steel, the endurance strength is approximately

$$S'_{e,\,steel} = 0.5S_{ut} \qquad [S_{ut} < 200{,}000 \text{ psi}]$$
$$= 100{,}000 \text{ psi} \qquad [S_{ut} > 200{,}000 \text{ psi}]$$

If necessary, the ultimate strength of steel can be calculated from its *Brinell hardness.*

$$S_{ut} \approx (500)(BHN)$$

FLUCTUATING STRESSES

For a part subjected to a fluctuating load, failure cannot be determined solely by yield strength or endurance limit. The combined effects of loading must be considered. The stress is described graphically on a diagram which plots the mean stress versus the alternating stress. Both of these stresses may be either normal stresses or shear stresses. A criterion for failure is established by relating the yield strength, the ultimate strength, and the endurance limit. One method of relating this information is a *Soderberg line.*

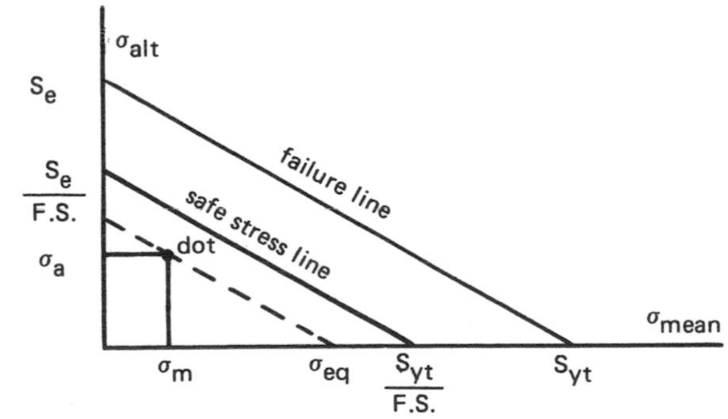

The *mean stress* is

$$\sigma_m = \frac{\sigma_{max} + \sigma_{min}}{2}$$

The *alternating stress* is half the *range stress.*

$$\sigma_a = \frac{\sigma_{max} - \sigma_{min}}{2}$$

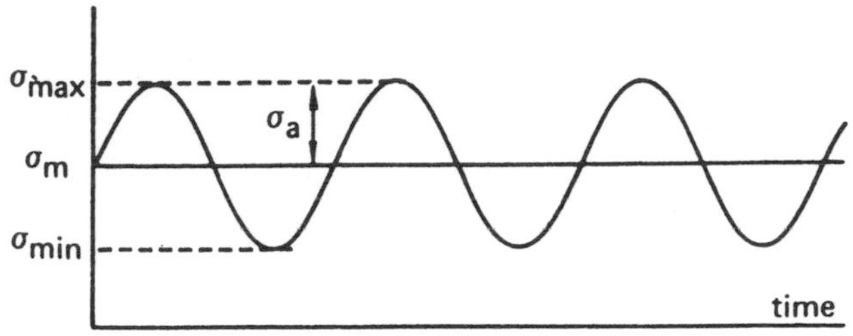

Stress concentration factors (with the notable exception of the Wahl correction factor for springs) are applied to the alternating stress only.

The factor of safety is

$$FS = \frac{S_{yt}}{\sigma_{eq}}$$

CUMULATIVE FATIGUE

The *Palmgren-Miner cycle ratio summation theory* (usually called just *Miner's rule*) often is used to evaluate cumulative damage. If a machine part is subjected to $\sigma_{max,\,1}$ for n_1 cycles, $\sigma_{max,\,2}$ for n_2 cycles, etc., the part should not fail if the equation holds. (N_i are the fatigue lives for the various stress levels.)

$$\sum \frac{n_i}{N_i} < C$$

A value of 1.0 commonly is used for *C*.

DYNAMICS

NEWTON'S LAWS

The first law: The velocity (momentum) of an object will not change unless it is acted upon by a force.

The second law: $F = \frac{ma}{g_c}$ or $T = I\frac{\alpha}{g_c}$

The third law: For every action there is an equal but opposite reaction.

Law of Universal Gravitation: The gravitational force between two objects with masses m_1 and m_2 is

$$F = \frac{Gm_1m_2}{d^2}$$

$$G = 3.44\ \text{EE}-8\ \frac{\text{ft}^4}{\text{lbf-sec}^4}$$

DISTANCE, VELOCITY, AND ACCELERATION

$$a = \frac{dv}{dt} = \frac{d^2s}{dt^2}$$

$$v = \frac{ds}{dt} = \int a\,dt$$

$$s = \int v\,dt = \iint a\,dt$$

UNIFORM ACCELERATION FORMULAS

to find	*given these*	*use this formula*
t	a, v_o, v	$t = (v - v_o)/a$
t	a, v_o, s	$t = \frac{\sqrt{2as + v_o^2} - v_o}{a}$
t	v_o, v, s	$t = 2s/(v_o + v)$
a	t, v_o, v	$a = (v - v_o)/t$
a	t, v_o, s	$a = (2s - 2v_ot)/t^2$
a	v_o, v, s	$a = (v^2 - v_o^2)/2s$
v_o	t, a, v	$v_o = v - at$
v_o	t, a, s	$v_o = (s/t) - \frac{1}{2}at$
v_o	a, v, s	$v_o = \sqrt{v^2 - 2as}$
v	t, a, v_o	$v = v_o + at$
v	a, v_o, s	$v = \sqrt{v_o^2 + 2as}$
s	t, a, v_o	$s = v_ot + \frac{1}{2}at^2$
s	a, v_o, v	$s = (v^2 - v_o^2)/2a$
s	t, v_o, v	$s = \frac{1}{2}t(v_o + v)$

ACCELERATION DUE TO GRAVITY

Falling body problems may be solved with the uniform acceleration formulas by substituting $a = g = 32.2$ ft/sec^2.

PROJECTILE MOTION

The formulas neglect air drag. Range (r) is maximum when $\phi = 45°$.

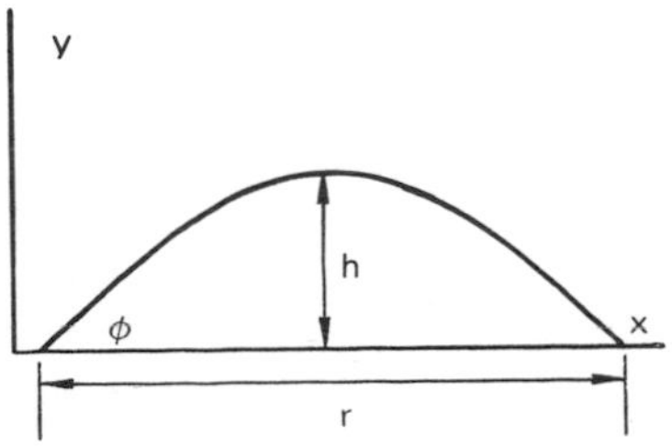

$$x = (v_o\cos\phi)t$$

$$y = (v_o\sin\phi)t - \frac{1}{2}gt^2$$

$$v = \sqrt{v_o^2 - 2gy}$$

$$v_x = v_o\cos\phi$$

$$v_y = v_o\sin\phi - gt$$

$$h = \frac{(v_o^2\sin^2\phi)}{2g}$$

$$r = \frac{(v_o^2\sin 2\phi)}{g}$$

$$T = \frac{(2v_o\sin\phi)}{g}$$

WORK AND ENERGY

Work is an energy transfer that occurs when a force is moved through a distance:

$$W = \int F \cdot ds$$

Potential energy is the energy that an object possesses by virtue of its position in a gravitational field:

$$E_p = \frac{mgh}{g_c} \quad \text{(in ft-lbf)}$$

For a spring (with rate constant k) compressed an amount x:

$$E_p = \frac{1}{2}kx^2$$

Kinetic energy is the energy that an object possesses by virtue of its velocity:

$$E_k = \frac{mv^2}{2g_c} \quad \text{(translation)}$$

$$E_k = \frac{I\omega^2}{2g_c} \quad \text{(rotation)}$$

THE WORK-ENERGY PRINCIPLE

In the absence of thermal changes, the work done on (or by) a system is equal to its change in energy.

$$W = \Delta E_p + \Delta E_k$$

Power is the amount of work done per unit time.

$$P = \frac{W}{\Delta t}$$

Power can be calculated from force and velocity.

$$P = Fv \quad \text{(translation)}$$

$$P = T\omega \quad \text{(rotation)}$$

HORSEPOWER REQUIRED TO MAINTAIN VELOCITY

For translation: $hp = Fv/550$ (v in fps)

For rotation: $hp = 2\pi Tn/33{,}000$ (T in ft-lb, n in rpm)

POWER CONVERSIONS

$$\text{horsepower} = \frac{\text{ft-lbf/sec}}{550} = \frac{\text{ft-lbf/min}}{33{,}000} = 1.34\ \text{(kw)}$$

$$\text{kilowatts} = \frac{\text{ft-lbf/sec}}{737} = \frac{\text{ft-lbf/min}}{44{,}220} = .746\ \text{(hp)}$$

$$\text{BTU/sec} = \frac{\text{ft-lbf/sec}}{778} = \frac{\text{ft-lbf/min}}{46{,}680} = .707\ \text{(kw)}$$

ROTATIONAL MOTION

Analogous variables for rotational motion:

α: rotational acceleration
ω: rotational speed
θ: rotational position

$$\alpha = \frac{d\omega}{dt} = \frac{d^2\theta}{dt^2}$$

$$\omega = \frac{d\theta}{dt} = \int \alpha \, dt$$

$$\theta = \int \omega \, dt = \int\int \alpha \, dt$$

ROTATIONAL MOMENT OF INERTIA

$$I = \frac{mr^2}{2} \qquad \text{(solid cylinder)}$$

$$I = \frac{m(r_i^2 + r_o^2)}{2} \qquad \text{(hollow cylinder)}$$

RELATIONSHIP BETWEEN ROTATIONAL AND LINEAR MOTION

$$s = r\theta$$

$$v = r\omega$$

$$a = r\alpha$$

CENTRIFUGAL FORCE

$$F_c = \frac{mv^2}{g_c r}$$

HIGHWAY BANKING

The banking angle (superelevation) needed on a circular curve of radius r is

$$\phi = \arctan\,[v^2/gr]$$

PARALLEL AXIS THEOREM

$$I_{new} = I_{centroidal} + m(d^2)$$

RADIUS OF GYRATION

$$k = \sqrt{I/m}$$

IMPULSE-MOMENTUM PRINCIPLE

$$F\Delta t = m\Delta v/g_c \qquad \text{(translation)}$$

$$T\Delta t = I\Delta\omega v/g_c \qquad \text{(rotation)}$$

IMPACTS

The coefficient of restitution (e) is 1 if the collision is perfectly elastic. e is 0 if the collision is completely inelastic (the bodies stick together).

$$e = \frac{\text{relative separation velocity}}{\text{relative approach velocity}} = \frac{-(v_2' - v_1')}{(v_2 - v_1)}$$

CONSERVATION OF MOMENTUM IN A COLLISION

$$m_1v_1 + m_2v_2 = m_1v_1' + m_2v_2'$$

If $e = 1$, kinetic energy is also conserved.

UNDAMPED, FREE VIBRATIONS (SIMPLE HARMONIC MOTION)

$$f = \frac{\omega}{2\pi} = \frac{1}{T}$$

$$T = \frac{1}{f} = \frac{2\pi}{\omega}$$

$$\omega = 2\pi f = \frac{2\pi}{T}$$

For a simple spring-mass system,

$$\omega = \sqrt{\frac{k}{m} g_c} = \sqrt{\frac{g}{\delta_{static}}}$$

For a simple pendulum consisting of a mass (m) at the end of a massless cord of length L,

$$\omega = \sqrt{\frac{g}{L}}$$

SHAFT VIBRATIONS (CRITICAL SPEEDS)

The critical speed of a shaft can be found from its free lateral vibration frequency. The general equation for the fundamental frequency (critical speed in revolutions per second) is

$$f = \frac{1}{2\pi}\sqrt{\frac{g\Sigma(m_i\delta_i)}{\Sigma(m_i\delta_i^2)}}$$

m_i is the mass of the ith rotating body, and δ_i is the static deflection of the shaft at the ith body.

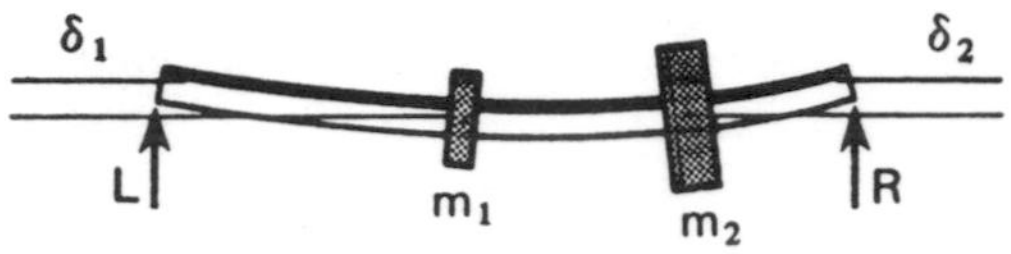

Since this equation is difficult to use, it is common to substitute the *Dunkerly approximation.*

$$\left(\frac{1}{f}\right)^2_{composite} = \left(\frac{1}{f_1}\right)^2 + \left(\frac{1}{f_2}\right)^2 + \left(\frac{1}{f_3}\right)^2 + \cdots$$

The classical solution to the problem of a shaft carrying a single object of weight W assumes a weightless shaft. The deflection at the load is

$$\delta = \frac{Wa^2b^2}{3EIL} \qquad \text{(simple supports)}$$

$$\delta = \frac{Wa^3b^3}{3EIL^3} \qquad \text{(fixed supports)}$$

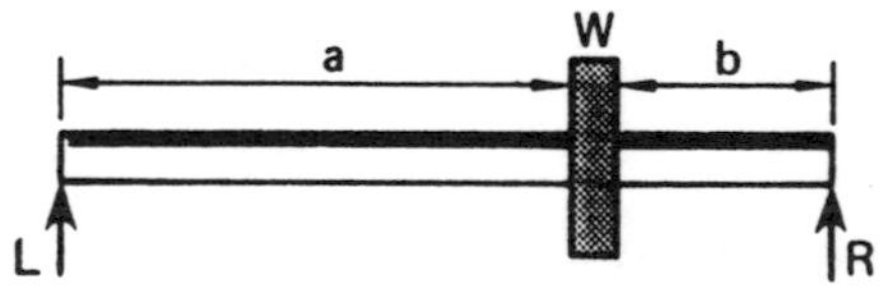

$$I = \frac{\pi r^4}{4} \qquad \text{(solid round shafts)}$$

$$f = \frac{1}{2\pi}\sqrt{\frac{g}{\delta}} \qquad \text{(consistent units)}$$

DAMPED, FREE VIBRATIONS

C is the damping coefficient, ω is the natural frequency.

$$C_{critical} = \frac{2m\omega}{g_c}$$

$$\text{damping ratio} = \frac{C}{C_{critical}} = \frac{n}{\omega}$$

$$n = \frac{Cg_c}{2m}$$

$n < \omega$ is underdamping

$n = \omega$ is critical damping (fastest return)

$n > \omega$ is overdamping

$$\omega_{\text{damped}} = \sqrt{\omega^2 - n^2}$$

$$= \omega \sqrt{1 - \left(\frac{C}{C_{\text{critical}}}\right)^2}$$

$$\text{logarithmic decrement} = \frac{2\pi n}{\omega_{\text{damped}}}$$

DAMPED, FORCED VIBRATIONS

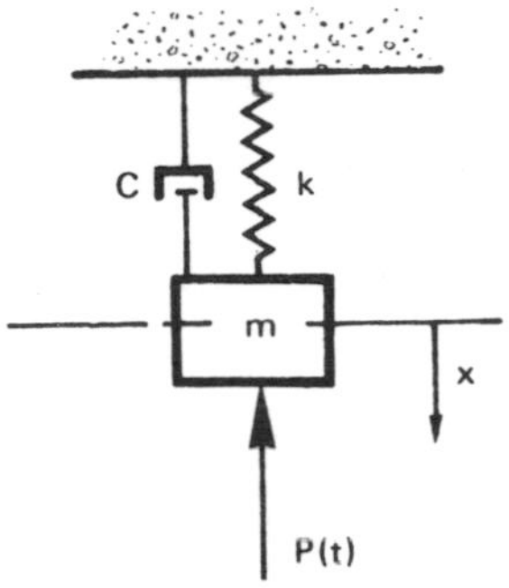

The amplitude of forced vibrations can be found from the pseudo-static deflection and the *magnification factor*, β.

$$A = \beta \left(\frac{P}{k}\right) = \beta \delta_{\text{static}}$$

$$\beta = \frac{1}{\sqrt{\left[1 - \left(\frac{\omega_f}{\omega}\right)^2\right]^2 + \left[\frac{g_c C \omega_f}{m(\omega)^2}\right]^2}}$$

$$= \frac{1}{\sqrt{\left[1 - \left(\frac{\omega_f}{\omega}\right)^2\right]^2 + \left[\frac{2C\omega_f}{C_{\text{crit}}\omega}\right]^2}}$$

If the forcing frequency, ω_f, is equal to the natural frequency, ω, the magnification factor will be very large. This condition is known as *resonance*.

VIBRATION ISOLATION AND CONTROL

$$f = \frac{1}{2\pi} \sqrt{\frac{g}{\delta_{st}}}$$

$$\sigma_{st} = \frac{\text{equipment weight}}{k}$$

The amount of isolation is known as the *isolation efficiency*.

$$\eta = 1 - \frac{1}{\left(\frac{f_f}{f}\right)^2 - 1}$$

Damping is beneficial only in the case of $(f_f/f) > \sqrt{2}$. Otherwise, damping is detrimental because the transmissibility actually increases above 1.0. The *transmissibility* is the ratio of transmitted force to unbalanced force. For undamped systems ($C = 0$), the transmissibility is equal to the magnification factor, β. For damped systems, the transmissibility is

$$TR = \frac{P_{\text{transmitted}}}{P_{\text{applied}}}$$

$$= \beta \sqrt{1 + \left[\frac{2C\omega}{C_{\text{crit}}\omega_f}\right]^2}$$

NOISE CONTROL

SOUND PRESSURE LEVEL

The minimum perceptible pressure amplitude has been standardized as 20 $\mu N/m^2$ (formerly expressed as 0.0002 μbar). Sound pressure level is

$$L_p = 10 \log_{10}\left[\frac{p}{20\text{ EE }-6}\right]^2 = 20 \log_{10}\left[\frac{p}{20\text{ EE }-6}\right]$$

SOUND POWER LEVEL

Sound power level, L_W, while not directly measurable, is used to describe the characteristics of a given noise source. Sound power level is a logarithmic power ratio with a standard reference level of EE−12 watts.

$$L_W = 10 \log_{10}\left[\frac{W}{\text{EE}-12}\right]$$

NOISE REDUCTION

Noise reduction (L_{NR}) is the difference in sound pressure due to modification of noise sources or the room.

$$L_{NR} = L_{P_o} - L_{P_m}$$

INSERTION LOSS

Insertion loss, L_{IL}, is used to describe the noise reduction of a noise control device installed between the source and the observer.

$$L_{IL} = L_{p_o} - L_{p_z} = 10 \log\left(\frac{p_z}{p_o}\right)^2 \qquad \text{for pressure}$$

$$= L_{W_o} - L_{W_z} = 10 \log\left(\frac{W_z}{W_o}\right) \qquad \text{for power}$$

TRANSMISSION LOSS

Transmission loss, L_{TL}, is related to the portion of sound energy transmitted through a barrier. The fraction of incident energy transmitted through a barrier is the *transmission coefficient*, τ.

$$L_{IL} = 10 \log \frac{1}{\tau}$$

ADDITION AND SUBTRACTION OF DECIBELS

Summation of decibels cannot be accomplished by simple addition because logarithmic ratios are involved.

$$L_{total} = 10 \log_{10} \sum 10^{(L_i/10)}$$

ROOM CONSTANT

Acoustic absorption characteristics of a room can be described by a *room constant, R*. A room with perfect absorbing surfaces (anechoic) has $R = \infty$, while a room with perfect reflecting surfaces (reverberant) has $R = 0$. The room constant is related to the average *absorption coefficient*, ($\bar{\alpha}$), and the total surface area, A.

$$R = \frac{\bar{\alpha}A}{1 - \bar{\alpha}}$$

If a room contains absorbing materials with known absorption coefficients, a good estimate of average absorption coefficient can be obtained.

$$\bar{\alpha} = \frac{\Sigma A_i \alpha_i}{\Sigma A_i} = \frac{\Sigma S_i}{\Sigma A_i} = \frac{S}{A_{total}}$$

The product, $A\alpha$, is known as the *surface absorption*, S, in sabins, where one sabin is one square foot of perfectly abosrbing surface.

The *Sabin equation* relates the absorption coefficient to the volume, the surface area, and the time required for a sound to decay to 60 dB below the original level.

$$\bar{\alpha} = \frac{0.049V}{t_{60}A}$$

NOISE REDUCTION COEFFICIENT

The *Noise Reduction Coefficient* rating, NRC, simplifies comparisons of materials. NRC is the arithmetic average of the four absorption coefficients measured at test frequencies of 250, 500, 1000, and 2000 Hz.

NOISE DOSE

Daily *noise dose* is the actual exposure duration divided by the permissible exposure duration. If an employee is exposed to one or more different levels during his shift, his daily noise dose can be calculated as the sum of partial noise doses.

$$D = \left(\frac{C_1}{t_1}\right) + \left(\frac{C_2}{t_2}\right) + \left(\frac{C_3}{t_3}\right) + \cdots$$

NRC AND α COEFFICIENTS

Materials	α Coefficients 125 Hz	250 Hz	500 Hz	1000 Hz	2000 Hz	4000 Hz	NRC
Brick, unglazed	.03	.03	.03	.04	.05	.07	.04
Brick, unglazed, painted	.01	.01	.02	.02	.02	.03	.02
Carpet							
1/8″ Pile Height	.05	.05	.10	.20	.30	.40	.16
1/4″ Pile Height	.05	.10	.15	.30	.50	.55	.26
3/16″ combined Pile & Foam	.05	.10	.10	.30	.40	.50	.23
5/16″ combined Pile & Foam	.05	.15	.30	.40	.50	.60	.34
Concrete Block, painted	.10	.05	.06	.07	.09	.08	.07
Fabrics							
Light velour, 10 oz. per sq. yd., hung straight, in contact with wall	.03	.04	.11	.17	.24	.35	.14
Medium velour, 14 oz. per sq. yd., draped to half area	.07	.31	.49	.75	.70	.60	.56
Heavy velour, 18 oz. per sq. yd., draped to half area	.14	.35	.55	.72	.70	.65	.62
Floors							
Concrete or Terrazzo	.01	.01	.01	.02	.02	.02	.02
Linoleum, asphalt, rubber or cork tile on concrete	.02	.03	.03	.03	.03	.02	.03
Wood	.15	.11	.10	.07	.06	.07	.09
Wood parquet in asphalt on concrete	.04	.04	.07	.06	.06	.07	.06
Glass							
1/4″, sealed, large panes	.05	.03	.02	.02	.03	.02	.03
24 oz., operable windows (in closed condition)	.10	.05	.04	.03	.03	.03	.04
Gypsum Board, 1/2″ nailed to 2 X 4's 16″ o.c., painted	.10	.08	.05	.03	.03	.03	.05
Marble or Glazed Tile	.01	.01	.01	.01	.02	.02	.01
Plaster, gypsum or lime, rough finish or lath	.02	.03	.04	.05	04	.03	.04
Same, with smooth finish	.02	.02	.03	.04	.04	03	.03
Hardwood Plywood paneling 1/4″ thick, Wood Frame	.58	.22	.07	.04	.03	.07	.09
Water Surface, as in a swimming pool	.01	.01	.01	.01	.02	.03	.01
Wood Roof Decking, tongue-and-groove cedar	.24	.19	.14	.08	.13	.10	.14
Air, Sabins per 1000 cubic feet @ 50% RH				.9	2.3	7.2	
Audience, seated, depending on spacing and upholstery of seats*	2.5–4.0	3.5–5.0	4.0–5.5	4.5–6.5	5.0–7.0	4.5–7.0	
Seats, heavily upholstered with fabric*	1.5–3.5	3.5–4.5	4.0–5.0	4.0–5.5	3.5–5.5	3.5–4.5	
Seats, heavily upholstered with leather, plastic, etc.*	2.5–3.5	3.0–4.5	3.0–4.0	2.0–4.0	1.5–3.5	1.0–3.0	
Seats, lightly upholstered with leather, plastic, etc.*			1.5–2.0				
Seats, wood veneer, no upholstery*	.15	.20	.25	.30	.50	.50	

*Values given are in sabins per person or unit of seating at the indicated frequency

NUCLEAR ENGINEERING

EINSTEIN'S EQUATION FOR BINDING ENERGY

$$E = mc^2 \quad \text{(joules)}$$
(m in kilograms)
(c = 3 EE8 m/s)

PARTICLE MASSES (CARBON SCALE)

name	symbol	mass (amu)
proton	${}_1H^1$ or p	1.007277
neutron	${}_0n^1$ or n	1.008665
electron	${}_{-1}e^0$ or β	.000548593
positron	${}_{+1}e^0$	.000548593
alpha particle	${}_2He^4$ or α	4.00260
gamma	γ	0.0000000
deuteron	${}_1H^2$ or D	2.01345
carbon	C	12.00000

RADIOACTIVE DECAY

Radioactive decay can be described mathematically through the use of a *decay constant*, λ, which is independent of the environment. The number of disintegrations per second is known as the *activity, A*, typically measured in curies. One curie is equal to 3.7 EE10 disintegrations per second (dps).

The decay constant can be determined from the *half-life*, $t_{1/2}$, or from the *mean life expectancy* of the atom, $\bar{t}$.

$$\lambda = \frac{.693}{t_{1/2}} = \frac{1}{\bar{t}}$$

The mass, activity, and number of atoms at time t can be calculated from the values at $t = 0$ once λ is known.

$$m_t = m_0 e^{-\lambda t}$$
$$A_t = A_0 e^{-\lambda t} = \lambda N_t$$
$$N_t = N_0 e^{-\lambda t}$$

ATOMS PER UNIT VOLUME

N_0 is Avogadro's number, 6.023 EE23 atoms/gmole.

$$N = \frac{\rho N_0}{(\text{A.W.})}$$

CROSS SECTIONS

The effective size of atoms is measured in square centimeters, analogous to the size of a target at which the neutrons are beamed. Since the effective area is small, *cross sections* (short for "cross sectional areas") usually are expressed in *barns*, where one barn equals EE−24 cm². The symbol for the cross section is σ. Neutrons can interact with atoms in two basic ways. They can be absorbed or scattered.

$$\sigma_t = \sigma_a + \sigma_s$$

It is useful to think of cross sections as being *interaction probabilities* per unit length of path. Thus, the probability of a neutron being absorbed by an atom is

$$p\{\text{absorption}\} = \frac{\sigma_a}{\sigma_t}$$

The probability that a neutron will interact after traveling distance x (in cm) through a substance is

$$p\{\text{interacting}\} = \sigma_t N x$$

If a neutron beam with incident density, J (neutrons/cm²-sec) hits a target of area, A, the average number of interactions per second will be

$$\text{interaction rate} = J A \sigma_t N x$$

MACROSCOPIC CROSS SECTIONS

The product $N\sigma$ is given the special symbol Σ and the special name, *macroscopic cross section.*

THERMAL NEUTRONS

Usually, the terms "thermal", "20°C", ".0253 eV", and "2200 m/sec" are considered synonymous.

The term *thermal* implies that the neutrons have experienced sufficient collisions to bring them down to the same energy level as that of the surrounding atoms, usually assumed to be at 20°C. Of course, not all neutrons travel at the same velocity. In fact, a Maxwellian velocity distribution similar to that assumed in the kinetic theory of gases occurs. Such a distribution predicts the *most probable particle velocity* to be

$$v_p = 1.28\text{ EE4}\sqrt{T} \quad \text{(cm/sec)}$$

Since the velocity is very low, the kinetic energy is

$$E_k = \tfrac{1}{2}mv^2$$

It is important to know the neutron energy level since cross sections vary with neutron energy. In the low-energy (less than resonant) region, called the *linear* or *(1/v) region*, the actual cross section can be found from the thermal data, which is widely tabulated. (E is in eV, T is in °K.)

$$\sigma' = \frac{\sigma_{th} V_{th}}{V} = \sigma_{th}\sqrt{\frac{293}{T}} = \sigma_{th}\sqrt{\frac{.0253}{E}}$$

APPROXIMATE NEUTRON THERMAL CROSS SECTIONS (σ_{th} in barns)

(es = elastic scatter, c = capture, f = fission, a = absorption, t = total)

	σ_{es}	σ_c	σ_f	σ_a	σ_t
U-natural*	8.3	3.43	4.16	7.59	15.89
U-235	10	101	577	678	688
U-238	8.3	2.8	.0005	2.8	11.1
U-233		48	525	573	
Pu-239	9.6	274	741	1015	1025
Pu-241		425	950	1375	

*Assumes composition is .72% U-235 and 99.28% U-238.

CROSS SECTION VALUES FOR REACTOR CALCULATIONS

The average value to be used in reactor calculations is obtained by assuming a Maxwellian distribution.

$$\bar{\sigma} = \frac{\sigma_{th}}{1.128}$$

If the (1/v) assumption is invalid, as it is for some moderators, uranium, and plutonium, a *non-1/v correction factor, g'*, must be used.

$$\bar{\sigma} = \frac{g'\sigma_{th}}{1.128}$$

Finally, if the reaction is not at 20°C, the average value of σ is

$$\bar{\sigma} = \frac{g'\sigma_{th}}{1.128}\sqrt{\frac{293}{T}} = \frac{g'\sigma_{th}}{1.128}\sqrt{\frac{.0253}{E}}$$

INTENSITY OF A POINT SOURCE

If a point source emits S rays or particles per second isotropically, the flux at a distance x without attenuation is given by

$$\phi(x) = \frac{S}{4\pi x^2}$$

BETA AND GAMMA ATTENUATION

Traditionally, the macroscopic cross section has been called the *attenuation coefficient* or the *linear absorption coefficient*.

$$\mu_l = \Sigma_t = \sigma_t N \qquad (1/\text{cm})$$

The *mass attenuation coefficient* (or the *mass absorption coefficient*) can be derived from the linear absorption coefficient.

$$\mu_m = \frac{\mu_l}{\rho} \qquad (\text{cm}^2/\text{g})$$

μ_l and μ_m are used to determine the attenuation of a monoenergetic narrow-beam gamma source passing through a shield of thickness, x. The units of ρx are g/cm², a typical measure of shield thickness.

$$I = I_0 e^{-\mu_l x} = I_0 e^{-\mu_m \rho x}$$

This intensity disregards all scattering and energy build-up, factors that usually must be considered.

The *half-value thickness* and the *half-value mass* will reduce the radiation beam 50%.

$$x_{1/2} = \frac{.693}{\mu_l}$$

$$m_{1/2} = \frac{.693}{\mu_m}$$

NEUTRON ATTENUATION

Thermal neutron intensity within a shield follows the exponential decay rule. x is measured along the path.

$$J = J_0 e^{-\Sigma_a x}$$

It is not possible to absorb fast neutrons.

IRRADIATION AND ACTIVATION

Substances form radioactive atoms when irradiated by a neutron beam. The *generation rate* per unit volume of activated sample is

$$g = \phi N \sigma_a \left(\frac{\text{atoms}}{\text{cm}^3\text{-sec}}\right)$$

The number of activated atoms at time t during the irradiation is

$$N_t = N_0 e^{-\lambda t} + \frac{(\text{volume})g}{\lambda}(1 - e^{-\lambda t})$$

$$= N_0 e^{-\lambda t} + \frac{N_s \sigma_a \phi}{\lambda}(1 - e^{-\lambda t})$$

N_0 is the original number of unstable atoms (if any). λ is the decay constant. N_s is the total number of atoms in the sample.

The activity at any time is

$$A = \lambda N_t$$

After irradiation stops at t_2,

$$N_t = N_{t2} e^{-\lambda t}$$

REACTOR POWER

The *reactor rate* depends on the average flux and is defined as $\Sigma_a \bar{\phi}$. The number of fissions per second per cubic centimeter of core is $\Sigma_f \bar{\phi}$. The *power density*, or *watt density*, is $\Sigma_f \bar{\phi}/3.1$ EE10.

The *thermal power* from a reactor is

$$P_{th} = \frac{\Sigma_f \bar{\phi} V}{3.1 \text{ EE}10}$$

$$\approx 3.92 \text{ EE} - 8\bar{\phi} m_f$$

P_{th} is in watts if the mass is in kilograms and the volume is in cubic centimeters.

In power calculations, the reactor power (*watts-thermal*) must be distinguished from the transmitted power (*watts-electric*). The transmitted power is

$$P_e \text{ (watts)} = \eta P_{th} \text{ (\% of reactor capacity)}$$

η typically is in the order of 40%.

DOUBLING TIME OF BREEDER REACTOR

An approximate formula for the *doubling time* (the time to double the number of plutonium atoms) given the original plutonium mass in grams, the reactor power in megawatts, the fuel consumption per unit power, m' (typically 1.23 to 1.3 grams/MW-day), and the conversion ratio (CR), is

$$\text{linear } t_d = \frac{m}{(\text{CR} - 1)(m')(P)} \qquad \text{(days)}$$

$$\text{exponential } t_d = (.693)(\text{linear } t_d) \qquad \text{(days)}$$

The *linear doubling time* assumes that all excess plutonium remains in the reactor. The *exponential doubling time* assumes that excess plutonium is removed regularly and is placed in a second breeder reactor.

The *conversion ratio*, CR, is the ratio of the number of plutonium atoms created to the number used. CR − 1 is the *breeding gain*. CR will be less than 1 for conversion operation and greater than 1 for breeding.

NUCLEAR CONVERSION FACTORS

1 amu	1.66053×10^{-24} g
	931.481 MeV
1 curie	3.7×10^{10} disintegrations/sec
1 coulomb	2.998×10^{9} esu
1 erg	1 g cm²/sec²
	0.6242×10^{6} MeV
1 eV	1.60210×10^{-12} erg
	1.517×10^{-22} BTU
	4.44×10^{-26} kW-hr
	1.60219 EE−19 joule
1 fission	~ 200 MeV (total)
	8.9×10^{-18} kW-hr (total)
	~ 180 MeV (in fuel)
1 joule	10^{7} ergs
1 hp	2545 BTU/hr
	0.7457 kW
	550 ft-lbf/sec
1 unit electronic charge	4.80×10^{-10} esu
1 watt	1 joule/sec
	3.4 BTU/hr

MODELING OF ENGINEERING SYSTEMS

RESPONSE VARIABLES

A dependent variable which predicts the performance of a system is known as a *response variable*. Position, velocity, and acceleration vary with time and are the dependent response variables. Time is usually the independent variable.

NATURAL, FORCED, AND TOTAL RESPONSE

Natural response is induced when energy is applied to an engineering system and subsequently is removed. The system is left alone and is allowed to do what it would do naturally, without the application of further disturbing forces. If a system is acted upon by a force which repeats at regular intervals, the system will move in accordance with that force. This is known as *forced response*. The natural and forced responses are present simultaneously in forced systems. The sum of the two responses is known as the *total response*.

FORCING FUNCTIONS

An equation which describes the introduction of energy into the system as a function of time is known as a *forcing function*. The most common forcing functions used in the analysis of engineering systems are the sine and cosine functions.

$$F(t) = \sin\omega t.$$

TRANSFER FUNCTIONS

The ratio of the system response (output) to the forcing function (input) is known as the *transfer function*, $T(t)$.

$$T(t) = \frac{R(t)}{F(t)}$$

Transfer functions generally are written in terms of the s variable. This is accomplished, if $T(t)$ is known, by taking the Laplace transform of the transfer function. The result is the *transform of the transfer function*, typically just called the *transfer function*.

$$T(s) = \pounds[T(t)]$$

PREDICTING SYSTEM RESPONSE

Assuming that $T(s)$ and $\pounds[F(t)]$ are known, the response function can be found by performing an inverse transformation.

$$R(t) = \pounds^{-1}[R(s)] = \pounds^{-1}[\pounds[F(t)]T(s)]$$

POLES AND ZEROS

A *pole* of the transfer function is a value of s which makes $T(s)$ infinite. Specifically, a pole is a value of s which makes the denominator of $T(s)$ zero. A *zero* of the transfer function makes the numerator of $T(s)$ zero. Poles and zeros can be real or complex quantities. Poles and zeros can be repeated within a given transfer function—they need not be unique.

A rectangular coordinate system based on the real-imaginary axes is known as an *s-plane*. If poles and zeros are plotted on the s-plane, the result is a *pole-zero diagram*. Poles are represented on the pole-zero diagram as ×'s. Zeros are represented as O's.

PREDICTING SYSTEM RESPONSE FROM POLE-ZERO DIAGRAMS

Poles on the pole-zero diagram can be used to predict the usual response of engineering systems. Zeros are not used.

- *pure oscillation:* Sinusoidal oscillation will occur if a pole-pair falls on the imaginary axis. A pole with a value of $\pm ja$ will produce oscillation with a natural frequency of $\omega = a$ radians/sec.
- *exponential decay:* Pure exponential decay is indicated when a pole falls on the real axis. A pole with a value of $-r$ will produce an exponential with time constant $(1/r)$.
- *damped oscillation:* Decaying sinusoids result from pole-pairs in the second and third quadrants of the s-plane. A pole-pair having the value $r \pm ja$ will produce oscillation with natural frequency of

$$\omega_n = \sqrt{r^2 + a^2}$$

The closer the poles are to the real axis, the greater will be the damping effect. The closer the poles are to the imaginary axis, the greater will be the oscillatory effect.

STABILITY

A pole with a value of $-r$ on the real axis corresponds to an exponential response of e^{-rt}. Similarly, a pole with a value of $+r$ on the real axis corresponds to an exponential response of e^{rt}. However, e^{rt} increases without limit. For that reason, such a pole is said to be unstable.

Since any pole in the first and fourth quadrants of the s-plane will correspond to a positive exponential, a stable system must have poles limited to the left half of the s-plane (i.e., quadrants two and three).

FEEDBACK THEORY

The basic feedback system consists of a *dynamic unit*, a *feedback element*, a *pick-off point*, and a *summing point*. The summing point is assumed to perform positive addition unless a minus sign is present.

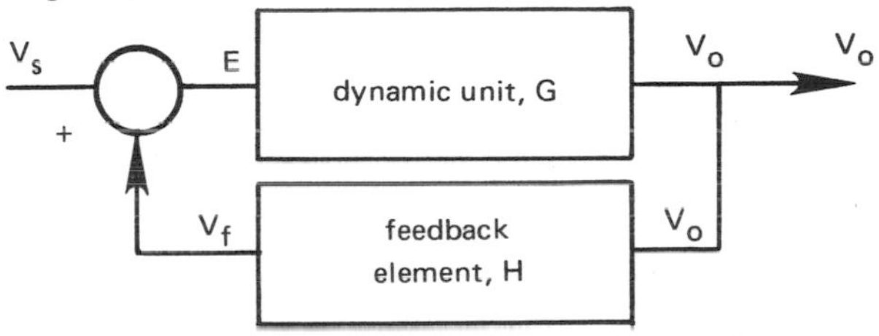

The dynamic unit transforms E into V_o according to the *forward transfer function*, *G*.

$$V_o = GE$$

For amplifiers, the forward transfer function is known as the *direct* or *forward gain*. The difference between the signal and the feedback is known as the *error*.

$$E = V_s + V_f$$

E/V_s is known as the *error ratio*. V_f/V_s is known as the *primary feedback ratio*.

The pick-off point transmits V_o back to the feedback element. The output of the dynamic unit is not reduced by the pick-off point. As the picked-off signal travels through the feedback loop, it is acted upon by the *feedback* or *reverse transfer function*, *H*.

The output of a feedback system is

$$V_o = GV_s + GHV_o$$

The *closed-loop transfer function* (also known as the *control ratio* or the *system function*) is the ratio of the output to the signal.

$$G_{loop} = \frac{V_o}{V_s} = \frac{G}{1 - GH}$$

SI SYSTEMS

SELECTED CONVERSION FACTORS TO SI UNITS

	SI Symbol	Multiplier to Convert From Existing Unit to SI Unit	Multiplier to Convert From SI Unit to Existing Unit
Area			
Circular Mil	µm²	506.7	0.001 974
Foot Squared	m²	0.092 9	10.764
Mile Squared	km²	2.590	0.386 1
Yard Squared	m²	0.836 1	1.196
Energy			
Btu (International)	kJ	1.055 1	0.947 8
Erg	µJ	0.1	10.0
Foot Pound-Force	J	1.355 8	0.737 6
Horsepower Hour	MJ	2.684 5	0.372 5
Kilowatt Hour	MJ	3.6	0.277 8
Meter Kilogram-Force	J	9.806 7	0.101 97
Therm	MJ	105.506	0.009 478
Kilogram Calorie (International)	kJ	4.186 8	0.238 8
Force			
Dyne	µN	10.	0.1
Kilogram-Force	N	9.806 7	0.101 97
Ounce-Force	N	0.278 0	3.597
Pound-Force	N	4.448 2	0.224 8
KIP	N	4 448.2	0.000 224 8
Heat			
Btu Per Hour	W	0.293 1	3.412 1
Btu Per (Square Foot Hour)	W/m²	3.154 6	0.317 0
Btu Per (Square Foot Hour °F)	W/(m²·°C)	5.678 3	0.176 1
Btu Inch Per (Square Foot Hour °F)	W/(m·°C)	0.144 2	6.933
Btu Per (Cubic Foot °F)	MJ/(m³·°C)	0.067 1	14.911
Btu Per (Pound °F)	J/(kg·°C)	4 186.8	0.000 238 8
Btu Per Cubic Foot	MJ/m³	0.037 3	26.839
Btu Per Pound	J/kg	2 326.	0.000 430
Length			
Angstrom	nm	0.1	10.0
Foot	m	0.304 8	3.280 8
Inch	mm	25.4	0.039 4
Mil	mm	0.025 4	39.370
Mile	km	1.609 3	0.621 4
Mile (International Nautical)	km	1.852	0.540
Micron	µm	1.0	1.0
Yard	m	0.914 4	1.093 6
Mass (weight)			
Grain	mg	64.799	0.015 4
Ounce (Avoirdupois)	g	28.350	0.035 3
Ounce (Troy)	g	31.103 5	0.032 15
Ton (short 2000 lb.)	kg	907.185	0.001 102
Ton (long 2240 lb.)	kg	1 016.047	0.000 984 2
Slug	kg	14.593 9	0.068 522
Pressure			
Bar	kPa	100.0	0.01
Inch of Water Column (20°C)	kPa	0.248 6	4.021 9
Inch of Mercury (20°C)	kPa	3.374 1	0.296 4
Kilogram-force per Centimeter Squared	kPa	98.067	0.010 2
Millimeters of Mercury (mm·Hg) (20°C)	kPa	0.132 84	7.528
Pounds Per Square Inch (P.S.I.)	kPa	6.894 8	0.145 0
Standard Atmosphere (760 torr)	kPa	101.325	0.009 869
Torr	kPa	0.133 32	7.500 6

SELECTED CONVERSION FACTORS TO SI UNITS (continued)

	SI Symbol	Multiplier to Convert From Existing Unit to SI Unit	Multiplier to Convert From SI Unit to Existing Unit
Power			
Btu (International) Per Hour	W	0.293 1	3.412 2
Foot Pound-Force Per Second	W	1.355 8	0.737 6
Horsepower	kW	0.745 7	1.341
Meter Kilogram-Force Per Second	W	9.806 7	0.101 97
Tons of Refrigeration	kW	3.517	0.284 3
Torque			
Kilogram-Force Meter (kg·m)	N·m	9.806 7	0.101.97
Pound-Force Foot	N·m	1.355 8	0.737 6
Pound-Force Inch	N·m	0.113 0	8.849 5
Gram-Force Centimeter	mN·m	0.098 067	10.197
Temperature			
Fahrenheit	°C	5/9 (°F − 32)	(9/5°C) + 32
Rankine	K	(°F + 459.67)5/9	(°C + 273.16)9/5
Velocity			
Foot Per Second	m/s	0.304 8	3.280 8
Mile Per Hour	m/s	0.447 04	2.236 9
	or	*or*	*or*
	km/h	1.609. 34	0.621 4
Viscosity			
Centipoise	mPa·s	1.0	1.0
Centistoke	μm²/s	1.0	1.0
Volume (Capacity)			
Cubic Foot	*l* (dm³)	28.316 8	0.035 31
Cubic Inch	cm³	16.387 1	0.061 02
Cubic Yard	m³	0.764 6	1.308
Gallon (U.S.)	*l*	3.785	0.264 2
Ounce (U.S. Fluid)	ml	29.574	0.033 8
Pint (U.S. Fluid)	*l*	0.473 2	2.113
Quart (U.S. Fluid)	*l*	0.946 4	1.056 7
Volume Flow (Gas-Air)			
Standard Cubic Foot Per Minute	m³/s	0.000 471 9	2119.
	or	*or*	*or*
	l/s	0.471 9	2.119
	or	*or*	*or*
	ml/s	471.947	0.002 119
Standard Cubic Foot Per Hour	ml/s	7.865 8	0.127 133
	or	*or*	*or*
	μ*l*/s	7 866.	0.000 127
Volume Liquid Flow			
Gallons Per Hour (U.S.)	*l*/s	0.001 052	951.02
Gallons Per Minute (U.S.)	*l*/s	0.063 09	15.850

MISCELLANEOUS SUBJECTS

STRAIN GAGES

A strain gage is a folded wire which exhibits a resistance change as the wire length changes. The resistance change will be small, and temperature effects should be compensated by using a second unstrained gage as part of the bridge measurement system. Nichrome wire with a total resistance under 1000 ohms commonly is used.

The *strain sensitivity (gage factor)* is defined as

$$K = \left(\frac{\Delta R}{R_o}\right)\left(\frac{L_o}{\Delta L}\right)$$

The strain (generally in microinches per inch) is related to the resistance change.

$$\epsilon = \frac{\Delta R}{KR_o}$$

TEMPERATURE-SENSITIVE RESISTORS

Resistance in conductors will vary with temperature. The change can be positive or negative, and the variation is nonlinear. The variation can be calculated if the coefficients of thermal resistance are known. (β generally is small and is not considered in most analyses).

$$R_T + R_o(1 + \alpha\Delta T + \beta\Delta T^2)$$
$$\Delta T = T - T_o$$

Coefficients of Thermal Resistance, α (1/°C)

conductors			
aluminum	0.0043	manganin	0.00002
brass	0.0020	nichrome	0.0004
constantan	0.000002	nickel	0.0068
copper	0.0067	platinum	0.0039
gold	0.0040	silver	0.0041
iron	0.0061	tin	0.0046
lead	0.0039		

ITEM RELIABILITY

Reliability as a function of time, $R(t)$, is the probability that an item will continue to operate satisfactorily up to time t. Reliability often is described by the *negative exponential distribution*. Specifically, it is assumed that an item's reliability is

$$R(t) = 1 - F(t) = e^{-\lambda t} = e^{-t/MTBF}$$

This infers that the probability of x failures in a period of time is given by the Poisson distribution.

$$p\{x\} = \frac{e^{-\lambda}\lambda^x}{x!}$$

The negative exponential distribution is appropriate whenever an item fails only by random causes but never experiences deterioration during its life. This implies that the *expected future life* of an item is independent of the previous duration of operation.

The *hazard function* is defined as the conditional probability of failure in the next time interval given that no failure has occurred thus far. For the exponential distribution, the hazard function is

$$z(t) = \lambda$$

Since this is not a function of t, exponential failure rates are not dependent on the length of time previously in operation.

SERIAL SYSTEM RELIABILITY

A serial system will fail if any one of the components fails. The system reliability is

$$R^* = R_1R_2R_3...R_n$$

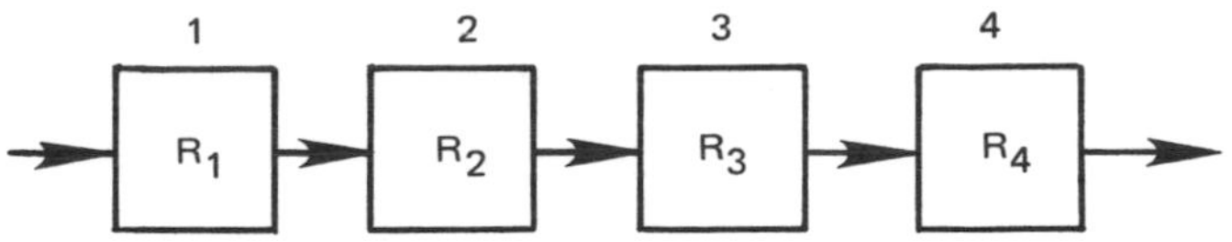

PARALLEL SYSTEM RELIABILITY

A parallel system with n items will fail only if all n items fail. This property is called *redundancy*, and such a system is said to be redundant.

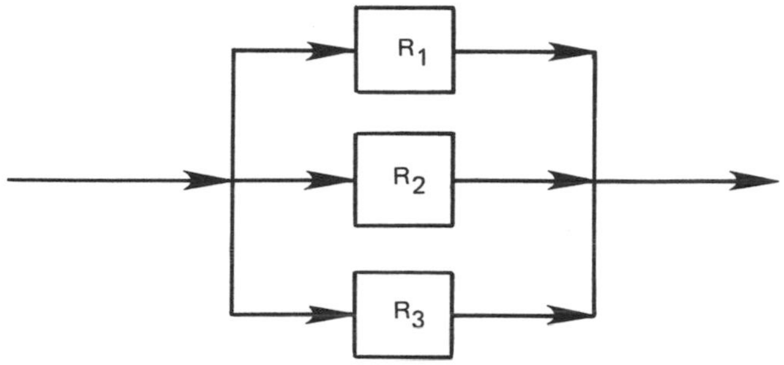

LEARNING CURVES

- T_1 time or cost for the first item
- T_n time or cost for the nth item
- n total number of items produced
- b learning curve constant

Learning Curve Constants

learning curve	b
75%	.415
80%	.322
85%	.234
90%	.152
95%	.074

The time to produce the nth item is given by

$$T_n = T_1(n)^{-b}$$

The total time to produce units from quantity n_1 to n_2 inclusive is

$$\frac{T_1}{(1-b)}\left[\left(n_2 + \frac{1}{2}\right)^{1-b} - \left(n_1 - \frac{1}{2}\right)^{1-b}\right]$$

ECONOMIC ORDER QUANTITY

The *economic order quantity* (EOQ) is the order quantity which minimizes the inventory costs per unit time.

- a the constant depletion rate $\left(\frac{\text{items}}{\text{unit time}}\right)$
- h the inventory storage cost $\left(\frac{\$}{\text{item-unit time}}\right)$
- H the total inventory storage cost between orders ($)
- K the fixed cost of placing an order ($)
- Q_0 the order quantity

PROFESSIONAL PUBLICATIONS INC. • P.O. Box 199, San Carlos, CA 94070

If the original quantity on hand is Q_0, the stock will be depleted at

$$t^* = \frac{Q_0}{a}$$

The total inventory storage cost between t_o and t^* is

$$H = \frac{1}{2}h\frac{Q_o^2}{a}$$

The total inventory and ordering cost per unit time is

$$C_t = \frac{aK}{Q_o} + \frac{1}{2}hQ_o$$

C_t can be minimized with respect to Q_o. The EOQ and time between orders are:

$$Q_o^* = \sqrt{2\frac{aK}{h}}$$

$$t^* = \frac{Q_o^*}{a}$$

Quick – I need additional study materials!

Please rush me the review materials I have checked. I understand any item may be returned for a full refund within 30 days. I have provided my bank card number as method of payment, and I authorize you to charge your current prices against my account.

	Solutions Manuals:
For the E-I-T Exam:	
[ ] Engineer-In-Training Review Manual	[ ]
[ ] Engineering Fundamentals Quick Reference Cards	
[ ] E-I-T Mini-Exams	
For the P.E. Exams:	
[ ] Civil Engineering Reference Manual	[ ]
[ ] Civil Engineering Sample Examination	
[ ] Civil Engineering Quick Reference Cards	
[ ] Seismic Design	
[ ] Timber Design	
[ ] Structural Engineering Practice Problem Manual	
[ ] Mechanical Engineering Reference Manual	[ ]
[ ] Mechanical Engineering Quick Reference Cards	
[ ] Mechanical Engineering Sample Examination	
[ ] Electrical Engineering Reference Manual	[ ]
[ ] Chemical Engineering Reference Manual	[ ]
[ ] Chemical Engineering Practice Exam Set	
[ ] Land Surveyor Reference Manual	[ ]
Recommended for all Exams:	
[ ] Expanded Interest Tables	
[ ] Engineering Law, Design Liability, and Professional Ethics	

SHIP TO:

Name ______________________________

Company ______________________________

Street ______________________ Apt. No. ________

City ____________________ State ______ Zip ________

Daytime phone number ______________________________

CHARGE TO (required for immediate processing):

______________________________ ________

VISA/MC/AMEX account number — expiration date

____

name on card

____

signature

Send more information

Please send me descriptions and prices of all available E-I-T and P.E. review books. I understand there will be no obligation.

____

____

____

____

A friend of mine is taking the exam too. Send additional literature to:

____

____

____

____

I disagree...

I think there is an error on page ________. Here is the way I think it should be.

Title of this book: **MECHANICAL ENGINEERING QUICK REFERENCE CARDS, fifth printing**

[ ] Please tell me if I am correct.

Contributed by (optional):

____

____

____

Affix
Postage

PROFESSIONAL PUBLICATIONS, INC.
1250 Fifth Avenue
Belmont, CA 94002

Affix
Postage

PROFESSIONAL PUBLICATIONS, INC.
1250 Fifth Avenue
Belmont, CA 94002

Affix
Postage

PROFESSIONAL PUBLICATIONS, INC.
1250 Fifth Avenue
Belmont, CA 94002